# Harnessing Digital Innovation for Air Transportation

Ayse Asli Yilmaz
*Atilim University, Turkey*

A volume in the Advances in Mechatronics and Mechanical Engineering (AMME) Book Series

IGI Global
PUBLISHER of TIMELY KNOWLEDGE

Published in the United States of America by
IGI Global
Engineering Science Reference (an imprint of IGI Global)
701 E. Chocolate Avenue
Hershey PA, USA 17033
Tel: 717-533-8845
Fax: 717-533-8661
E-mail: cust@igi-global.com
Web site: http://www.igi-global.com



Library of Congress Cataloging-in-Publication Data

Names: Yilmaz, Ayşe, 1981- editor.
Title: Harnessing digital innovation for air transportation / edited by: Ayse Yilmaz.
Description: Hershey PA : Engineering Science Reference, [2024] | Includes bibliographical references. | Summary: "This book provides comprehensive insights, strategies, and practical guidance for effectively managing aviation operations, safety, security, and customer experience in the digital era"-- Provided by publisher.
Identifiers: LCCN 2023050868 (print) | LCCN 2023050869 (ebook) | ISBN 9798369307328 (hardcover) | ISBN 9798369307335 (ebook)
Subjects: LCSH: Aircraft industry--Technological innovations. | Aircraft industry--Data processing.
Classification: LCC TL724 .H37 2024 (print) | LCC TL724 (ebook) | DDC 387.7068--dc23/eng/20240424
LC record available at https://lccn.loc.gov/2023050868
LC ebook record available at https://lccn.loc.gov/2023050869

This book is published in the IGI Global book series Advances in Mechatronics and Mechanical Engineering (AMME) (ISSN: 2328-8205; eISSN: 2328-823X)

British Cataloguing in Publication Data
A Cataloguing in Publication record for this book is available from the British Library.

# Advances in Mechatronics and Mechanical Engineering (AMME) Book Series

ISSN:2328-8205
EISSN:2328-823X

Editor-in-Chief: J. Paulo Davim, University of Aveiro, Portugal

**MISSION**

With its aid in the creation of smartphones, cars, medical imaging devices, and manufacturing tools, the mechatronics engineering field is in high demand. Mechatronics aims to combine the principles of mechanical, computer, and electrical engineering together to bridge the gap of communication between the different disciplines.

The **Advances in Mechatronics and Mechanical Engineering (AMME) Book Series** provides innovative research and practical developments in the field of mechatronics and mechanical engineering. This series covers a wide variety of application areas in electrical engineering, mechanical engineering, computer and software engineering; essential for academics, practitioners, researchers, and industry leaders.

**COVERAGE**

- Medical Robotics
- Control Systems Modelling and Analysis
- Biologically Inspired Robotics
- Computer-Based Manufacturing
- Sustainable and green manufacturing
- Intelligent Sensing
- Design and Manufacture
- Mechanisms and machines
- Computational Mechanics
- Autonomous Systems

IGI Global is currently accepting manuscripts for publication within this series. To submit a proposal for a volume in this series, please contact our Acquisition Editors at Acquisitions@igi-global.com or visit: http://www.igi-global.com/publish/.

The Advances in Mechatronics and Mechanical Engineering (AMME) Book Series (ISSN 2328-8205) is published by IGI Global, 701 E. Chocolate Avenue, Hershey, PA 17033-1240, USA, www.igi-global.com. This series is composed of titles available for purchase individually; each title is edited to be contextually exclusive from any other title within the series. For pricing and ordering information please visit http://www.igi-global.com/book-series/advances-mechatronics-mechanical-engineering/73808. Postmaster: Send all address changes to above address. 

## Titles in this Series

*For a list of additional titles in this series, please visit:*
*http://www.igi-global.com/book-series/advances-mechatronics-mechanical-engineering/73808*

***Studies on Single and Double Actuator Based DC Attraction Type Levitation Systems Optimization Techniques***
Rupam Bhaduri (Dayananda Sagar University, India)
Engineering Science Reference • © 2023 • 231pp • H/C (ISBN: 9781668473887) • US $225.00

***Advances in MEMS and Microfluidic Systems***
Rajeev Kumar Singh (Department of Mechanical Engineering, Amity University, Noida, India) Rakesh Kumar Phanden (Department of Mechanical Engineering, Amity University, Noida, India) Basant Singh Sikarwar (Department of Mechanical Engineering, Amity University, Noida, India) and J. Paulo Davim (Department of Mechanical Engineering, University of Aveiro, Portugal)
Engineering Science Reference • © 2023 • 331pp • H/C (ISBN: 9781668469521) • US $240.00

***Structural Mathematical Modeling Applications in Technological Machines and Transportation Vehicles***
Andrey Eliseev (Irkutsk State Transport University, Russia & Irkutsk National Research Technical University, Russia)
Engineering Science Reference • © 2023 • 288pp • H/C (ISBN: 9781668472378) • US $225.00

***Handbook of Research on Emerging Designs and Applications for Microwave and Millimeter Wave Circuits***
Jamal Zbitou (LABTIC, ENSA of Tangier, University of Abdelmalek Essaadi, Morocco) Mostafa Hefnawi (Royal Military College of Canada, Canada) Fouad Aytouna (LABTIC, ENSA of Tetouan, University of Abdelmalek Essaadi, Morocco) and Ahmed El Oualkadi (LabTIC, ENSA of Tangier, University of Abdelmalek Essaadi, Morocco)
Engineering Science Reference • © 2023 • 542pp • H/C (ISBN: 9781668459553) • US $325.00

*For an entire list of titles in this series, please visit:*
*http://www.igi-global.com/book-series/advances-mechatronics-mechanical-engineering/73808*

IGI Global
PUBLISHER of TIMELY KNOWLEDGE

701 East Chocolate Avenue, Hershey, PA 17033, USA
Tel: 717-533-8845 x100 • Fax: 717-533-8661
E-Mail: cust@igi-global.com • www.igi-global.com

# Table of Contents

# Detailed Table of Contents



Digital assistants work with the support of artificial intelligence-supported systems and automation technologies. Due to the conveniences provided by digital assistants, they provide great benefits in many areas such as ensuring flight control in aviation operations, air traffic management, pilot training, and facilitating passenger services. Errors in flight operations and passenger experiences are reduced to a minimum with features such as voice commands transmitted through virtual assistants, artificial intelligence-based guidance system and automatic warnings. This book chapter has been created to cover the development of digital assistants in aviation and their impact on practice, with particular emphasis on key subtopics. In the content of the book chapter, the application of digital assistants in flight operations and passenger experiences in the field of aviation and future projects are evaluated based on current literature.



This book chapter offers an examination of the transformative influence of Artificial Intelligence (AI) within the aviation sector, focusing specifically on its application in predictive maintenance for enhancing operational efficiency. Through the investigation of current research and industry trends, and the utilization of a literature review as its methodological framework, this chapter elucidates the transformative impact of AI-driven predictive maintenance strategies on aviation operations. It explores how AI algorithms analyze vast amounts of data to predict potential equipment failures, enabling proactive maintenance interventions that minimize downtime and optimize

# Preface

Welcome to *Harnessing Digital Innovation for Air Transportation*, a comprehensive exploration of the transformative power of digitalization in the aviation industry. Authored by esteemed scholars and practitioners from leading academic institutions and industry organizations, this book focuses deeply into the multifaceted ways in which digital innovation is reshaping the landscape of air transportation.

In an era characterized by rapid technological advancement, the aviation industry finds itself at the forefront of digital transformation. With the integration of cutting-edge technologies such as artificial intelligence, data analytics, and cybersecurity measures, airlines, airports, and aviation stakeholders are embracing digital innovations to navigate an increasingly complex and competitive environment.

*Harnessing Digital Innovation for Air Transportation* serves as a comprehensive exploration of the intersection between digital innovation and the aviation industry. In recent years, the aviation sector has witnessed a profound transformation driven by advancements in technology, ushering in an era of digitalization that has revolutionized various facets of air transportation. At the heart of this transformation lies the integration of cutting-edge technologies such as artificial intelligence, data analytics, and cybersecurity measures. These technologies have not only enhanced the efficiency and safety of aviation operations but have also reshaped the way airlines, airports, and aviation stakeholders interact with passengers and manage their operations.

The subject matter of this book encompasses a wide range of topics, each offering valuable insights into different aspects of digital innovation in air transportation. From the utilization of digital assistants for flight control and passenger services to the adoption of predictive maintenance systems and the mitigation of cyber threats, each chapter provides a unique perspective on the evolving landscape of aviation in the digital age. Furthermore, this book explores the strategic implications of digitalization for various stakeholders within the aviation industry. Whether it's examining the role of social media influencers in shaping public relations strategies or exploring the synergies between tourism and aviation in delivering seamless travel experiences, each chapter offers practical insights and analysis that are relevant to industry professionals, researchers, and policymakers alike.

Overall, *Harnessing Digital Innovation for Air Transportation* aims to provide a comprehensive overview of the transformative impact of digitalization on the aviation industry. By examining key trends, challenges, and opportunities, this book seeks to inform and inspire stakeholders to embrace digital innovation and leverage its potential to shape the future of air transportation. In today's rapidly evolving world, the topic of digital innovation in air transportation holds significant relevance and importance. The aviation industry serves as a critical global infrastructure, facilitating commerce, tourism, and connectivity on a scale never seen before. As society becomes increasingly interconnected and reliant on air travel, the need for innovative solutions to enhance safety, efficiency, and passenger experience has never been more pressing.

At the same time, technological advancements, particularly in the fields of artificial intelligence, data analytics, and cybersecurity, are driving profound changes across industries. In this context, the aviation industry stands out as a prime candidate for digital transformation, given its complex operational environment, stringent safety requirements, and high levels of customer expectations. The integration of digital technologies into air transportation has the potential to unlock numerous benefits and opportunities. From improving aircraft maintenance procedures through predictive analytics to optimizing air traffic management systems using AI algorithms, digital innovation promises to enhance operational efficiency, reduce costs, and minimize environmental impact.

Moreover, in an era where customer experience is paramount, digitalization offers airlines and airports new ways to engage with passengers and deliver personalized services. From mobile check-in and self-service kiosks to in-flight entertainment systems and social media engagement strategies, digital technologies are reshaping every touchpoint of the passenger journey. Furthermore, cybersecurity has emerged as a critical concern in aviation, given the industry's reliance on interconnected systems and digital infrastructure. As cyber threats become more sophisticated and pervasive, ensuring the resilience and security of aviation networks and data systems is essential to safeguarding the integrity of air transportation. In light of these developments, the topic of digital innovation in air transportation occupies a central place in the global discourse on technology, transportation, and sustainability. By exploring the latest trends, challenges, and opportunities in this field, stakeholders across the aviation ecosystem can stay informed, adapt to change, and harness the transformative power of digitalization to shape a more efficient, safe, and sustainable future for air transportation.

*Harnessing Digital Innovation for Air Transportation* caters to a diverse audience of professionals, researchers, policymakers, and enthusiasts who are interested in understanding and leveraging the transformative potential of digitalization in the aviation industry. The target audience for this book includes, but is not limited to, the following groups:

***Aviation Professionals:*** Pilots, air traffic controllers, aircraft maintenance engineers, airport operators, and other aviation professionals seeking to stay abreast of the latest developments in digital technologies and their applications in air transportation. This audience will benefit from insights into how digital innovation is reshaping operational procedures, safety protocols, and customer experiences within the aviation sector.

***Researchers and Academics:*** Scholars, researchers, and students studying fields such as aerospace engineering, computer science, aviation management, and transportation systems will find valuable insights and empirical evidence in this book. The diverse perspectives presented in each chapter offer opportunities for further exploration and academic inquiry into the impacts of digitalization on air transportation.

***Industry Stakeholders:*** Executives, managers, and decision-makers within airlines, airports, aviation technology companies, and regulatory agencies will gain strategic insights into the opportunities and challenges posed by digital innovation in air transportation. This audience can leverage the knowledge and best practices shared in the book to inform their organizational strategies, investments, and partnerships in the evolving digital landscape.

***Policymakers and Regulators:*** Government officials, policymakers, and regulatory authorities responsible for shaping aviation policies, standards, and regulations will find this book to be a valuable resource for understanding the implications of digitalization on aviation safety, security, and sustainability. The insights provided can inform policy decisions and regulatory frameworks aimed at promoting innovation while ensuring the integrity and resilience of the aviation ecosystem.

***Enthusiasts and General Readers:*** Enthusiasts of aviation, technology, and innovation who have a keen interest in understanding the intersection of digitalization and air transportation will find this book to be an engaging and informative read. Whether passionate about the future of flight or curious about the latest advancements in aviation technology, this audience will appreciate the accessible, yet insightful discussions presented in the book.

Overall, *Harnessing Digital Innovation for Air Transportation* is designed to appeal to a wide range of audiences with varying levels of expertise and interests in the aviation industry. By offering a comprehensive overview of digital innovation in air transportation, this book aims to inform, inspire, and empower stakeholders to embrace and leverage the opportunities presented by digitalization for the advancement of air transportation.

Embark on a journey through the dynamic realm of digital innovation in air transportation as it is focused into the rich tapestry of insights offered by each chapter in *Harnessing Digital Innovation for Air Transportation*.

## Chapter 1: Cleared for Take-Off: Exploring Digital Assistants in Aviation

This chapter illuminates the revolutionary impact of digital assistants powered by artificial intelligence (AI) on various aspects of aviation operations. By leveraging AI-supported systems and automation technologies, digital assistants contribute significantly to enhancing flight control, air traffic management, pilot training, and passenger services. Through a thorough examination of key subtopics, this chapter provides insights into the development of digital assistants in aviation and evaluates their practical implications based on current literature. By shedding light on the transformative potential of digital assistants, this chapter underscores their role in optimizing safety, efficiency, and passenger experience in the aviation industry.

## Chapter 2: Emerging Role of Artificial Intelligence in Aviation: Using Predictive Maintenance for Operational Efficiency

This chapter delves into the transformative potential of artificial intelligence (AI) and data analytics in optimizing maintenance processes and improving operational efficiency within the aviation industry. Through a comprehensive literature review and methodological approach, the chapter explores key applications of AI-driven predictive maintenance, intelligent flight operations, and passenger experience enhancements. By highlighting the significance of AI from a user perspective, this chapter contributes valuable insights to the existing body of knowledge, paving the way for ongoing dialogue and collaboration in this rapidly evolving field.

## Chapter 3: Aviation Cybersecurity: Shielding Against Cyberthreats in the Air

Focusing on cybersecurity practices specialized for aviation systems, this chapter emphasizes the critical importance of safeguarding against cyber threats in the air. By providing a comprehensive discussion of cybersecurity approaches and applications in the aviation sector, the chapter underscores the need for dynamic, totalistic, and adaptive cybersecurity measures. With heightened sensitivities and complexities in the aviation sector, a concrete understanding and approach to cybersecurity are essential for ensuring the integrity and resilience of aviation networks and data systems.

## Chapter 4: How Social Media Influencers Play a Role in the Aviation Industry by Online Engagement

This chapter explores the evolving role of social media influencers in shaping aviation public relations and passenger engagement strategies. By examining the effectiveness of online engagement through social media platforms, the chapter highlights the strategic significance of leveraging social media influencers in the aviation industry. Through an analysis of two-way symmetric communication processes and user-generated content, this chapter provides insights into enhancing communication activities and fostering meaningful interactions with target audiences.

## Chapter 5: The Pros and Cons of Digitalization in Aviation: Navigating the Digital's Potential Risks into Opportunities

In this chapter, the authors provide a nuanced analysis of the benefits and challenges associated with digital transformation in aviation. By navigating the potential risks and opportunities of digitalization, the chapter offers valuable perspectives on capitalizing on digital transformation while mitigating its associated risks. Through discussions on the latest digital trends and the role of digital transformation in shaping the future of aviation, this chapter aims to contribute novel perspectives to the ongoing discourse on digitalization in aviation management.

## Chapter 6: The Synergy Between Tourism, Aviation, and Digital Transformation: A Path to Seamless Travel Experiences

This chapter explores the profound impact of digital transformation on the tourism and aviation sectors, highlighting the necessity for a comprehensive digital strategy in response to global changes. By examining the interconnectedness of tourism and aviation, the chapter emphasizes the role of digital strategies in enhancing customer satisfaction, operational efficiency, and industry competitiveness. Through an exploration of innovations, increased connectivity, and affordability, this chapter offers insights into fostering tourism growth and shaping seamless travel experiences through digital transformation.

## Chapter 7: The Synergy between Tourism, Aviation, and Digital Transformation: A Path to Seamless Travel Experiences

This chapter serves as a beacon illuminating the profound impact of digital transformation on both the tourism and aviation sectors. In this chapter, the necessity for a comprehensive digital strategy in response to global changes is underscored, highlighting how technological advancements such as mobile applications, social media, and digital platforms are revolutionizing customer satisfaction, experience, and overall competitiveness within the tourism industry. Furthermore, the chapter explores how digital developments within the aviation sector, including smart maintenance, risk management, air traffic optimization, and customer relations, are crucial in enhancing operational efficiency and passenger experience. By delving into the historical evolution of the symbiotic relationship between tourism and aviation, the chapter emphasizes the pivotal role of innovations, increased connectivity, and affordability in fostering tourism growth. Through its comprehensive analysis and insights, this chapter contributes to a deeper understanding of the interconnectedness between tourism, aviation, and digital transformation, paving the way for seamless travel experiences in the digital age.

*Harnessing Digital Innovation for Air Transportation* offers a comprehensive and insightful exploration of the transformative power of digitalization in the aviation industry. Through a diverse range of chapters, the book covers key aspects of digital innovation, including the utilization of digital assistants, the role of artificial intelligence, cybersecurity measures, social media engagement, and the impact of digital transformation on tourism and aviation sectors. By providing valuable insights and perspectives from experts in the field, this book contributes to advancing knowledge and understanding of digitalization's impact on air transportation. It serves as a valuable resource for professionals, researchers, policymakers, and enthusiasts alike, inspiring innovative approaches to shaping the future of air travel in an increasingly digital world.

In conclusion, *Harnessing Digital Innovation for Air Transportation* significantly impacts the field by providing a comprehensive and insightful examination of the transformative potential of digitalization in aviation. Through its diverse array of chapters, the book offers valuable insights into key aspects of digital innovation, including the utilization of digital assistants, the integration of artificial intelligence, cybersecurity measures, social media engagement, and the synergies between tourism and aviation in the digital age. By bringing together contributions from experts in the field, the book contributes to advancing knowledge and understanding of digitalization's impact on air transportation, offering practical insights and recommendations for industry professionals, researchers, policymakers, and enthusiasts. By inspiring innovative approaches and fostering ongoing dialogue and collaboration, *Harnessing Digital Innovation for Air Transportation* serves as a valuable resource for navigating the evolving landscape of air travel in an increasingly digital world.

# Chapter 1
# Cleared for Takeoff:
## Exploring Digital Assistants in Aviation

**Seda Ceken**
https://orcid.org/0000-0002-5870-2246
*Istanbul University, Turkey*

## ABSTRACT

*Digital assistants work with the support of artificial intelligence-supported systems and automation technologies. Due to the conveniences provided by digital assistants, they provide great benefits in many areas such as ensuring flight control in aviation operations, air traffic management, pilot training, and facilitating passenger services. Errors in flight operations and passenger experiences are reduced to a minimum with features such as voice commands transmitted through virtual assistants, artificial intelligence-based guidance system and automatic warnings. This book chapter has been created to cover the development of digital assistants in aviation and their impact on practice, with particular emphasis on key subtopics. In the content of the book chapter, the application of digital assistants in flight operations and passenger experiences in the field of aviation and future projects are evaluated based on current literature.*

## INTRODUCTION

In today's world, many technologies can do much more than just create a communication and information flow for users. Developments in this field aim to create machines that mimic human interaction and emotional intelligence, to achieve rational, logic-based outcomes (Krakovsky, 2018). Artificial intelligence technologies are expected to play an exceedingly

DOI: 10.4018/979-8-3693-0732-8.ch001

crucial role in the future of the aviation sector. Investments in artificial intelligence, which amounted to approximately $340 million in 2019, are projected to reach $3.7 billion by 2027, with a compound annual growth rate of 45.3%. This growth rate indicates a significant shift in the aviation industry due to digital transformation. The migration of business processes to digital platforms enables the development of systems that provide real-time information flow to personnel, facilitating the easy completion of challenging and complex tasks (Innova, 2023). Additionally, according to a report published by Precedence Research, the global artificial intelligence in the aviation market was estimated to be $653.74 million in 2021. It is anticipated to exceed $9,985.86 million by 2030, demonstrating a registered Compound Annual Growth Rate (CAGR) of 35.38% from 2022 to 2030 (Precedence Research, 2022). These figures indicate that the global AI in the aviation market is projected to experience a substantial increase in the forecast period, driven by the growing adoption of big data analytics within the aerospace industry.

Airlines are using AI-based machines to improve passenger services, route planning and predictive maintenance to gain an edge in today's competitive environment. For example, companies using chatbots or virtual assistants can provide real-time information and personalized travel recommendations to passengers. Artificial Intelligence (AI) systems, with their revolutionary applications, lead to the beginning of a new process in the aviation world, improving operational efficiency, safety, security and passenger satisfaction. Digital and smart machines perform operations that require human intelligence by using artificial intelligence, digital assistants and smart maintenance systems in aviation systems. Among the conveniences provided by artificial intelligence are features such as automation in aviation operations, optimization of decisions and foreseeing maintenance needs that may arise. With these conveniences, it is frequently used by aviation companies, airports and aircraft manufacturers. Aviation companies that want to gain an advantage in the competitive environment can interact with passengers through chatbots or virtual assistants and provide real-time information and personalized travel recommendations. Artificial Intelligence (AI) is ushering in a transformative era in the aviation industry; it increases operational efficiency, safety and passenger experiences. While virtual assistants facilitate passenger services and flight operations efficiency, artificial intelligence-supported systems improve safety protocols based on flight operations and smart maintenance systems. (Jadoon, 2023). One of the most common applications of this technology is digital assistants, or, as they are known today, *voice-controlled virtual assistants, smart virtual assistants*, or *intelligent personal assistants.* Virtual assistants driven by AI technology are engineered to execute tasks and furnish information via voice or text-based interactions. In the aviation sector, they play a pivotal role in optimizing operational efficiency, enhancing the passenger journey, and raising safety standards. These highly intelligent systems exhibit versatility, proficiently managing a wide spectrum of functions, from addressing routine inquiries to executing intricate tasks with remarkable precision.

The aviation industry is experiencing rapid development by taking advantage of the opportunities offered by the latest technological developments. The source of this profound change is digital assistants. This book chapter has been created to address the development of digital assistants in aviation and their impact on practice, with particular emphasis on key sub-topics. In the section titled 'The Evolution of Digital Assistants in Aviation', how these technological developments have profoundly shaped the aviation industry is examined. Subsequently, under 'The Role of Digital Assistants in Flight Operations,' the pivotal functions undertaken by digital assistants for flight crews are subjected to meticulous scrutiny. 'Enhancing Passenger Experience with Digital Assistants' is aimed at assessing how this technology can contribute to the provision of a more favorable passenger experience. Furthermore, the chapter delves into an exploration of the 'Safety and Security: Contributions of Digital Assistants in Aviation' and a comprehensive analysis of the 'Digital Technologies in Aviation: Advantages and Challenges', thereby offering a holistic perspective on these critical aspects. Finally, within the segment labeled 'The Future of Digital Assistant Technology in the Aviation Industry', the potential future contributions of digital assistants within the aviation landscape are pondered.

## Digital Assistants

AI-based digital assistants, often referred to as virtual agents or virtual assistants, are autonomous entities designed to act on behalf of users within computer-based environments. Prominent examples include Apple's Siri, Microsoft's Cortana, and Amazon's Alexa. These digital assistants represent a significant advancement beyond earlier chatbot versions like ELIZA and A.L.I.C.E., thanks to progress in machine learning, microprocessing, natural language processing, and speech recognition. These systems utilize inputs, such as the user's voice, visual data (images), and contextual information, to offer support to users by responding to inquiries in a natural language, offering suggestions, and executing tasks (Hauswald et al., 2015).

The term *personal digital assistant* was first used by Alan Kay, a graduate student at the University of Utah in the 1970s. Kay's concept involved an interactive computer with wireless communication capabilities and a flat screen. The device he described was named the "*Dynabook,*" and its initial prototype was developed at the Palo Alto Research Center (PARC) in the 1970s. However, it was never further developed for commercial release. Nevertheless, Kay's idea served as a catalyst for the development of laptop computers and PDAs. In the early 1980s, Day-Timers, primarily used by the business community to track names, contact numbers, appointments, and important dates, began transitioning from paper-based systems to digital formats (Olcay, 2019). While researchers worldwide had been developing machines capable of comprehending and responding to speech for various straightforward tasks, the first AI (artificial intelligence) assistant introduced to a group of users in the United States was Wildfire. Wildfire was a telephone assistant created by Wildfire Communications, a Boston-based company that secured funding in 1992. It aimed to address specific user requirements by merging the convenience of a customizable directory service with the capabilities of an intelligent answering machine (Pierraccini, 2021). Since the 1990s, along with advancements in artificial intelligence technology, not only in algorithms but also in areas such as speech interpretation, natural language processing, and memory capacity, personal assistants have evolved significantly from their early iterations. Significant advancements in the realm of virtual assistants emerged in 2011 when Apple introduced Siri, the first smartphone-based virtual assistant, as part of its new iPhone 4S. Siri was developed from a spinoff of the Stanford Research Institute (SRI), which Apple had acquired a few years earlier. Its core technology relied heavily on the same speech recognition systems found in customer care telephone applications. Following Siri's success and with similar functionality, Microsoft unveiled Cortana in April 2014 as a virtual assistant available for their operating systems and devices. Cortana drew from Microsoft's extensive expertise in speech technology, nurtured through years of advanced research in the field. In September 2014, Apple introduced the *'Hey Siri'* feature, allowing users to activate the virtual assistant with a voice command without the need to press the home button on an iPhone or iPad. Just two months later, Amazon announced Alexa along with the Echo device (Pierraccini, 2021). These tools have now transformed into intelligent personal assistants (IPA), capable of not only performing specific tasks but also interacting with the user by analyzing their requests, providing responses, and even learning from them. Recent developments in machine learning and natural language processing have made voice the primary method for human-computer interaction. This development has given rise to service providers known as voice-activated digital assistants, which utilize speech recognition, speech synthesis, and natural language processing (Ramos, 2018).

## Evolution of Digital Assistants in Aviation

The rise of digital platforms is reshaping all sectors of the economy; each passing day, more industries are adapting their platforms to manage their businesses, find new partners, manage quality, enhance the performance of marketing and customer support functions. Digital transformation is widely recognized as a disruptive force that has been evolving for decades, fundamentally altering business paradigms, processes, and models (Léopold, 2018). In the continually growing realm of digital transformation initiatives, applications such as digital flight route optimization, predictive maintenance, and aircraft systems optimization are gaining prominence each day. In the dynamic landscape of the aviation industry, the evolutionary journey of digital assistants has been a testament to the relentless pursuit of innovation and technological advancement. The role of digital assistants in aviation has changed with artificial intelligence-supported systems depending on changing technologies and passenger demands. These changes also change the way professionals employed in the field of aviation use technology. Keeping up with the changes has not only brought changes in aviation operations but also the training and development of individuals using this technology.

Deregulation movements in the United States in 1978 and the shift of low-exchange businesses in the 1980s resulted in changes in the management and financing of airports operated as public utilities. As a result of this process, airports were privatized through tenders and became commercial centers (Jaffer & Timbrell, 2014). The Gulf War, the emergence of low-cost carriers in the 1990s, the events of September 11, 2001, and the global economic crisis of 2008 led airlines to place greater emphasis on cost efficiency. In particular, low-cost airline companies such as Ryanair decided to make radical changes and adopt the low-cost business model in order to survive. They undertook a comprehensive review of their operations, realigned processes, and product offerings, and leveraged the efficiency and expanded customer reach offered by the widespread availability of the Internet, all without the need for substantial investments in sales infrastructure (Silling, 2019). At this point, the focus has shifted towards passengers, who are the primary source of revenue for airports and their subsidiaries, and their wants and needs. In commercialized airports, the importance of both revenues and customers has been recognized over time, leading to a reduction in dependency on aviation-related revenues and a shift towards non-aviation revenues. By keeping aviation fees and taxes low, demand has been stimulated, thereby attracting traffic and passengers to airports and attempting to create alternative revenue streams (Cam & Durmaz, 2018).

## Role of Digital Assistants in Aviation Operations

Digital technologies play a crucial role in enhancing consumer satisfaction within the aviation industry, ultimately fostering greater customer loyalty. Airports can streamline the passenger journey from entrance to aircraft, improving operational safety. These technologies facilitate the efficient management of increasing passenger traffic, ensuring uninterrupted operations. Additionally, airports can employ digital tools for virtual simulations during peak hours to allocate resources effectively and optimize runway usage (Molchanova, 2020). The three most significant developments in the aviation sector concerning digital transformation are big data, the Internet of Things (IoT), artificial intelligence, and machine learning. Big data, especially in conjunction with the advancement of smart aviation, plays a critical role in various aviation applications, such as route planning, air traffic management, flight operations, aircraft design, performance improvement, and aviation business management. Analyzing the acquired data with artificial intelligence support leads to insights that help minimize flight risks. IoT sensors are utilized for data collection in aircraft engines and other components. These sensors play a crucial role in detecting potential maintenance issues before they pose significant risks. IoT enables the measurement and analysis of aircraft performance, resulting in reduced maintenance costs and increased operational efficiency. Artificial intelligence assists in managing operations within the aviation sector. According to data from the International Air Transport Association (International Air Transport Association [IATA], 2016), it is expected that the total number of passengers will exceed 7 billion by 2035. In response to the ever-increasing number of passengers, many airlines are strengthening their air traffic control systems with AI-based technologies (Innova, 2023). Research activities in the aviation sector have increased for the application of deep learning[1], from aircraft delay prediction to maintenance and repair. For instance, the use of big data, especially in unmanned platforms, becomes critical for accident prevention. In this regard, the large data flow resulting from communication between aircraft and other resources, for example, can be analyzed in real-time to adjust the collision avoidance trajectory in the air. With the availability of larger storage capacities in avionics systems, a sophisticated collision prevention policy among airborne aircraft can be effectively established (Çukur, 2020).

Currently utilized digital communication assistants in aviation provide convenience in aviation operations. In the context of call center operations and customer surveys, full automation is employed to streamline processes. This is achieved by converting traditional questionnaires into human-like dialogues and utilizing automated calls to systematically gather customer feedback responses. Additionally, the transition from conducting customer satisfaction surveys via phone calls to utilizing SMS, messenger, or email helps reduce the workload on service workshops, as these repetitive tasks can now be automated. Digital assistants, such as chatbots integrated into a company's communication platforms like *Slack* or *Teams*, play a pivotal role in enhancing employee productivity. They effectively handle tasks that would otherwise consume employees' time, including activities like scheduling meetings, coordinating participants, or parking reservations. Additionally, chatbots are employed as information systems for employees, aiding in the distribution of internal messages and offering substantial support in answering frequently asked questions (FAQs), thereby reducing the need for human involvement in these aspects (CloudFlight, n.d.).

The operation of certain digital assistants used in aviation is as follows:

1. **Mission+:** Mission+ is a Electronic Flight Assistant developed by NAVBLUE, offering pilots a comprehensive application that combines electronic navigation charts, mission management, and globe-centric weather information. This solution is designed to streamline the entire flight process, from pre-flight preparation to post-flight closure. For pilots utilizing Flysmart+, Airbus's aircraft performance calculations solution, the integration with Mission+ offers a convenient and efficient way to access Flysmart+ directly from the Mission+ platform. This seamless connection enables the automatic exchange of aircraft and flight data between Mission+ and Flysmart+, resulting in time savings and a reduction in the need for manual data entry, which in turn decreases the risk of errors (Airbus Aircraft, 2021; NavBlue, n.d.).
2. **AvioFlex:** The AvioFlex product family, developed by Innova, addresses the needs of institutions operating in the aviation sector with the latest technological solutions. Comprising five different solutions, namely *AvioFlex Cabin+, AvioFlex Crew Duty, AvioFlex Crew Check-in, AvioFlex LIFUS Manager*, and *AvioFlex Ops Manager*, the AvioFlex product family aims to enhance the operational efficiency and safety of businesses in the aviation industry. *ioFlex Cabin+ increases the efficiency of airline companies by assisting cabin crew in their duties. AvioFlex Cabin+,* the pre-flight, flight take-off and post-flight phases that apply to its hat, make many tasks waiting for cabin chiefs very manageable. During these stages, separate forms are prepared and information is given to cabin chiefs for all flight-related transactions. Many tasks are carried out, from material supply to performance management. Cabin+ allows easy storage of all sizes via tablet. *AvioFlex Crew Duty* offers a mobile platform that enables communication between pilots, cabin crew and other staff, allowing them to access all necessary details about flights via a web portal or their own mobile devices. The system greatly improves the flow of information between cabin and ground crews and helps manage personnel operations as required. *AvioFlex Crew Check-in* allows flight crews to check in with just a few taps on their mobile devices and view all necessary flight information. It provides significant time savings at airports, especially during

periods when air traffic is busy. Thanks to this application, airline companies benefit from human resources much more efficiently, preventing loss of time and preventing potential risks. *AvioFlex LIFUS* Manager facilitates the tracking of processes such as which training the instructors gave to which pilot or flight personnel and the notes taken by the relevant personnel during all training processes. With the use of tablets, processes are managed in real time, increasing the efficiency of airline companies. *The AvioFlex Ops Master* system provides airlines with ease in the management of their operation control centers. It has been specifically developed to enable them to cope with the ever-increasing number of flights, aircraft and destinations. This digital assistant replaces e-mails and all operational approval systems and offers the opportunity to track all flight operations from a single center with the authorization and reporting system that regulates the workflow of all business units. (Innova, 2023).

3. **MITRE - Digital Copilot:** The Digital Copilot is a software-based tool designed to enhance the experience of single-pilot operators in general aviation aircraft. It operates on a tablet device and incorporates proprietary algorithms and a unique user interface, serving as a cognitive assistant for pilots. This technology accesses a vast array of aviation information sources to monitor flight progress and provide safety alerts. Its potential applications include improving airspace awareness, ensuring clearance conformance, automating approach briefings, enhancing surface safety, depicting holding patterns, facilitating traffic pattern entries, and assisting with various other tasks (MITRE, n.d.).
4. **Victor5:** *Victor5* is a service designed to provide contingency and situation awareness support for air traffic control operations in cases of service disruption that necessitate a switch to ATC Zero or continued operations in a degraded mode. It primarily addresses scenarios where various Air Traffic Control (ATC) sub-systems may fail, including en route, approach, or tower operations. In some instances, the situation may require operational personnel to relocate from the primary ACC or TWR building to a secondary facility, which may or may not have back-up systems in place. Victor5 enables operational personnel to maintain situational awareness during transitions to backup equipment and ensures the safe management of

degraded operations, particularly in cases where backup equipment is unavailable due to common failures with primary equipment, such as radar feeds (DataBeacon, n.d.; Perez, 2021).

5. **EHang:** *EHang*, a Chinese company, and Airbus are collaborating to develop AI-based navigation technology. EHang incorporates AI in its autonomous aircraft, while Airbus has successfully executed taxiing, take-off, and landing maneuvers using vision-based AI. This increasing emphasis on adopting AI for various aviation operations is a key driver for the growth of the global AI in the aviation market (Precedence Research, 2022).

## Enhancing Passenger Experience With Digital Assistants

Airports are heavily investing in cutting-edge technologies to enhance the passenger experience and alleviate the challenges caused by workforce issues. Self-service technologies and biometrics are at the forefront of this transformation. Airports are particularly prioritizing self-service options like self-check-in and self-bag-drop, with an impressive 86% planning to implement these services by 2025. Furthermore, the adoption of a secure, single biometric token across all airport touchpoints has seen a significant increase, rising from 3% in 2021 to 39% in 2022, with more than half of airports planning to implement it in the next three years. This reflects a strong commitment to usher in the next generation of travel experiences, where passengers can effortlessly use their facial biometrics as their boarding pass, streamlining their journey through the airport (Airports Council International, 2023). Airlines gather vast amounts of information about their customers. They have begun to smartly process data from various sources and actions, such as reservations, itineraries, accommodations, queries, transportation, pricing, cancellations, customer feedback, geographic location, and more. Additionally, they have recognized the effectiveness of social media in enhancing passenger loyalty. Airlines, which were once cautious about social sharing platforms like Facebook, Twitter, and Instagram, have now discovered ways to use these channels effectively. They have found ways to analyze passengers' social status and economic status in detail and leverage this information to their advantage. In the present day, airlines possess valuable data about every passenger, ranging from personal information to travel habits, specific preferences (e.g. *Zoey*[2] assists travelers at their destinations. The process entails companies offering *Zoey*'s services to their travelers, who must register on the platform. Once registered, *Zoey* gathers travelers' preferences, provides travel advice, and addresses inquiries in a manner similar to a traditional travel agent), in-flight food and beverage choices, hotel and car rental habits, and more (Sanal Savunma, 2021).

Airlines and flight/mile/point programs use the following data to monitor, measure, and analyze passengers' behavior. These data are collected and analyzed by aviation companies to provide passengers with better services (Chang et al., 2019):

- *Search history*
- *Payment history*
- *Baggage reclaim knowledge*
- *Departure and arrival dates and times*
- *City of arrival and departure*
- *In-flight sales and spending in airport stores*
- *Food preferences on the plane*
- *Hotel and car rental connections*
- *Passenger numbers flow together*
- *Earned credit card points*
- *Miles earned and used*
- *Customers' influence on others on social media.*

Airlines are shifting towards a passenger-centric approach to address the needs of connected travelers. Across all channels, including travel agencies, airline products and ancillary businesses, are becoming more differentiated (Çukur, 2020). The demands related to air travel, such as accessibility to services and airports, speed, cleanliness, and convenience at transit points, have evolved and diversified over time. Choices that enhance today's customer experience and set it apart from the past are increasingly technology-driven. This is because the growing adaptation to and reliance on technology has made it inevitable to make current and future passenger experiences technology-centric. In the simplest terms, the prevalence of technology in mobile devices, the rise of social media, and the proliferation of wireless networks necessitate the design of terminals with a technology-centered approach. Such devices are no longer used consciously but have become a natural habit, indicating passengers' readiness and willingness to embrace technology-focused airports (Harrison et al., 2014).

The International Air Transport Association (IATA) announced that it is in the final stages of development for a mobile application called the "*IATA Travel Pass*," which will enable travelers to easily access COVID-19 identity information at airports through brochures. The Digital Travel Pass will link a person's test results, vaccine proof (when available), and identity verification to an electronic copy of the user's passport. According to IATA, it will serve as a resource listing various rules for nearby laboratory locations that meet the testing and/or vaccination requirements of national entry and destination points (IVisa, 2023).

Commercial airline companies use various applications to enhance passengers' travel experiences and adapt to the digital world. For instance, Delta Airlines has reduced costs and wait times by offering passengers self-service options such as mobile apps, kiosks, and baggage tracking. British Airways employs data-driven initiatives like the "*Know Me*" program to recognize individual customer needs and utilizes e-commerce and social media to expand its reach. Lufthansa introduces innovations such as self-check-in kiosks, mobile apps, and automated baggage systems, offering personalized flight packages through the "*Flightpass*" program based on individual preferences and travel patterns. Investments in e-commerce and social media help the airline reach new customers (Digital Adoption, 2023). In addition to these developments, Turkish Airlines provides passengers with digital services through the *Boti* digital travel assistant, including features like flight queries, flight price checks, flight status inquiries, check-in assistance, mobile boarding passes, airport maps, and promotional inquiries (Turkish Airlines Terminal, n.d.). Similarly, Pegasus Airlines offers similar passenger services through the *FlyBot* virtual assistant (Pegasus Airlines, n.d.).

Consequently, numerous airports worldwide have started investing in mobile applications, self-service check-in, bag drop, and way-finding maps to enhance the passenger experience. However, these technology-based investments are motivated not only by passengers but also by the increasing and forthcoming automation, which will reduce passenger processing costs and times. This reduction is a crucial factor in ensuring the projected growth in passenger traffic on a global scale, as building ever-larger airports or increasing the number of service personnel is economically, financially, operationally, and environmentally unsustainable. The only way to reconcile this contradiction is by creating more technology-focused passenger experiences.

## Digital Assistants in Airport Operations

The integration of digital assistants into airport operations is a significant development in the aviation industry and the beginning of a new era in which passenger experiences and operational efficiency increase. Powered by artificial intelligence (AI) and advanced technologies, virtual assistants are reinventing various aspects of airport management. They function as virtual guides that provide real-time information, simplify navigation, and even offer personalized recommendations to travelers. Additionally, virtual assistants are effective in streamlining administrative tasks, improving security protocols, and optimizing overall airport management. This shift toward digitalization underscores the industry's commitment to meeting evolving passenger demands and addressing the increasingly complex challenges posed by modern air travel. Airports are places where time is of the essence, so it is crucial to manage all of these operational processes efficiently and access data promptly. In this context, examples of the digital airport concept and marketing approaches include technological advancements that enable electronic and virtual shopping at airports like Dubai, Qatar, and Singapore Changi. Additionally, Frankfurt Airport has kiosks that exclusively provide services through QR codes. Changi, Schiphol, Heathrow, Incheon, Munich, Zurich, and Copenhagen airports are among others that have initiated and implemented digital transformation processes. At Dusseldorf Airport, robots are there to welcome you (UTED, 2019). The following paragraph provides information about digital assistants at some airports.

*Philos* is a virtual assistant deployed at Athens International Airport, designed to assist travelers. Powered by artificial intelligence, Philos offers information about travel-related matters, including Covid-19, and can communicate in both Greek and English. Philos engages with airport users through the airport's website, providing interactive and efficient responses to their inquiries before or after their flights. Since its launch in April 2021, Philos has successfully addressed over 120,000 questions, with 63% of these queries being made outside regular working hours (SmartRep, 2023). In another example, London Luton Airport made history as the first airport worldwide to adopt the *Tensator* Virtual Assistant technology developed by the Tensator Group. This innovative technology, designed for passenger interaction, provides a comprehensive solution to integrate marketing campaigns effectively during the point of sale. Through the use of imagery, music, and messaging that aligns with the broader campaign, it enhances awareness and reinforces the core message. Moreover, the Tensator Virtual Assistant offers a distinctive and captivating approach to delivering informational and instructional content. It animates messages, prompting individuals to stop and take notice, thus influencing changes in behavior (Tensator, 2023). An additional example of a digital assistant at Istanbul Airport is *Ask Me*. Positioned in various areas of the airport, Ask Me digital kiosks enable passengers to connect with live assistants via audio or video calls in English or Turkish, around the clock. These kiosks also offer accessibility for hearing-impaired passengers through sign language assistance. They assist travelers in exploring airport services, making inquiries about available amenities, locating service points and determining their proximity, generating routes from their current positions to relevant service points, scanning flight information, and identifying gate locations if announced (IGA Sustainability Report, 2019). In parallel with these developments, Frankfurt Airport, operated by Fraport, has initiated a pilot program in collaboration with Deutsche Bahn's dedicated IT service provider, DB Systel, to test a robot concierge named '*FRAnny*' (derived from the airport code 'FRA' for Frankfurt). FRAnny offers travelers information about directions to specific shops or restaurants, guidance to the correct gate, and assistance in accessing the airport's Wi-Fi. This digital customer service system is powered by artificial intelligence and a cloud-based voice-user interface (VUI), which can be employed in various forms, including chatbots, voice assistants, and robots. Developed by a team of Deutsche Bahn IT experts, FRAnny utilizes data from the airport's information system to comprehend and address travel-related queries and inquiries about airport facilities. Furthermore, FRAnny is multilingual, communicating in German, English, and seven other languages (The Moodie Davitt Report, 2019). Lastly, Munich Airport, incorporating digital assistants, offers passengers an enhanced digital experience as of October 4, 2023. Passengers can manually input their flight information or scan their boarding passes using tablets on portable baggage trolleys. This process provides access to real-time flight details and updates. Furthermore, passengers can select various services, such as shops and restaurants, and they receive location-based offers while being guided throughout the terminal with the assistance of an integrated map and directory. The usage of these smart baggage trolleys is complimentary and can be obtained in Terminal 2 after passing through the security checkpoint and prior to reaching the duty-free area (Munich Airport, 2023).

## Safety and Security: Contributions of Digital Assistants in Aviation

AI significantly enhances situational awareness and threat detection in airport security. It uses real-time video analysis to identify security risks, unauthorized personnel, and individuals on watchlists. AI-powered surveillance also detects anomalies like unattended baggage and overcrowding, allowing for immediate risk mitigation and passenger safety (Jadoon, 2023). Deep learning algorithms have the potential to amalgamate a wide range of passenger information, constructing a comprehensive risk profile. This profile may encompass travel history, passport details, visa, and biometric data, as well as flight behavior patterns. Airports can also establish connections with regulatory bodies, government agencies, and third-party data providers to compile a comprehensive risk profile for individuals. Risk-based screening at integrated networked checkpoints has the potential to adapt security measures in response to changing threat levels. This adaptation may involve using different detection algorithms based on factors such as the passenger's destination or their individual risk assessment. Additionally, it could include automatically rechecking the luggage of high-risk passengers. It is essential to ensure the confidentiality of information, adherence to privacy standards, secure data storage, and non-discriminatory use of this technology (Thompson, 2018). Aviation applications typically require higher levels of security compared to other fields. Therefore, when designing and developing a specialized big data system for aviation, enhanced security measures should be implemented. This can be an inhibiting factor when conducting human factor research activities. Since big data necessitates a broader information and system network rather than bringing together a few systems, as interpretation and management systems become more automated, issues may arise in ensuring the reliability of data (Cook & Rivas, 2016).

The primary purpose of AI digital assistants in aviation is to assist pilots during flight upset conditions, situations where a critical event occurs, and the flight crew has limited time to identify the issue and regain control of the aircraft. Incidents like AF 447[3] and the two B737 Max accidents[4], among others, fall into this category. While it may take flight crews precious minutes to understand the problem and determine corrective actions, a digital assistant, connected to various sensors and equipped with an extensive database of potential scenarios and recovery procedures, could potentially diagnose and respond to such situations in a matter of seconds. Therefore, the implementation of AI digital assistants in the cockpit is a top priority in aviation safety research. In addition, a Digital Assistant, particularly when interacting with a human crew, theoretically has the capability to record in much greater detail the thought process and decision-making during an event (whether recovered or not) than is currently achievable. In terms of safety, another contribution is that a Digital Assistant can serve as an extensive knowledge repository for aviation flight crews, controllers, and ground personnel with respect to procedures and potential hazards. When integrated into real-time operations involving multiple teams, it can offer immediate safety guidance concerning factors such as weather conditions and challenges that other teams might face. Moreover, this type of Digital Assistant can also retain daily NOTAMs (Notices to Airmen) and other briefings, disseminating them to the teams and ensuring that critical details are not overlooked during subsequent operations (Kirwan, 2023). Furthermore, predictive maintenance, facilitated by predictive analytics, secure remote connectivity, advanced sensors, and data analytics technology, significantly enhances the performance, reliability, and uptime of screening equipment. Through automated data transmission and predictive maintenance algorithms, AI transforms extensive data into practical insights for predictive service and performance. This enables airport operations teams to preemptively assess equipment condition and performance, allowing for proactive scheduling of system downtime to prevent failures (Thompson, 2018).When connected to live operations and multiple teams, digital assistants can provide real-time safety recommendations pertaining to factors such as weather conditions and challenges that other teams may encounter. This capability can enhance safety by facilitating swift responses and alerts during operations. All of these features can contribute to the improvement of safety in the aviation sector and the making of better decisions.

## Digital Technologies in Aviation: Advantages and Challenges

Digital assistants offer more and more advantages to flight crews and passengers in aviation operations. Machines with artificial intelligence systems can help minimize the amount of data. Due to improve flight performance, AI helps pilots analyze sensitive data during flight, such as fuel system, system status, weather and other important elements. The many sensors used in modern aircraft capture broader and more detailed data from the aircraft's many sources and systems. This data is used to monitor flight conditions and perform performance analysis after the trip. Pilots can leverage artificial intelligence (AI) assistants to make decisions using weather data and flight data transmitted from sensors. In severe weather conditions, these AI assistants can assist pilots with alternative routes. Artificial intelligence can be used to detect when two planes are on the same path, helping air traffic controllers issue warnings by telling pilots to reduce altitude. In addition, air traffic controllers can also monitor aircraft leaving or approaching the runway using artificial intelligence and smart cameras. Another advantage of digital assistants used in aviation is for tall passengers. Airlines offer self-check-in options for passengers through their websites. In order for passengers to have a smooth experience at airports, robots that serve as personal assistants are placed at certain points. In some airports, customers can buy their own boarding tickets using the machines available there. All of this is made possible through artificial intelligence-supported digital assistants. Additionally, AI plays a pivotal role in enabling the *Aircraft Real-time Health Monitoring System (ARHMS)*. Sensors are strategically placed throughout the aircraft to detect variations in parameters, such as pressure, temperature, magnetic force, and humidity. The data from these sensors are processed, and upon arrival, the entire aircraft undergoes a comprehensive scan and testing through ARHMS, thus preventing unforeseen catastrophic events. Moreover, one of the major inconveniences in air travel is the lengthy check-in lines at airports, where passengers often endure extended waiting times. Recognizing the significance of customer satisfaction and service quality in commercial aviation, airlines are turning to AI as a solution to enhance customer engagement and provide exceptional service. Airlines now offer self-check-in options through their websites, and in select airports, passenger-assisting robots are stationed at key locations to streamline the airport experience. Some airports even offer self-service boarding ticket kiosks, all made possible through the implementation of AI technology. Lastly, AI in the cockpit aids pilots in optimizing their flight path by analyzing and issuing alerts regarding weather conditions, fuel levels, and other pertinent factors (Abubakar et al., 2022).

The most significant barriers to the use of big data analytics in the airline industry are the insufficient time, resources, skills, tools (including those for forecasting and management purposes), and systems required to extract value from the data. From a data integration perspective, an ever-increasing number of data sources, such as aircraft sensors, radars, cameras, meteorological stations, and airports, generate highly dynamic data. For a given application, the size of data sets and the volume of data generated, analyzed, and interpreted per second can be exceptionally high. Data sets may exhibit heterogeneous formats and types, with situations where terabytes of data are processed per second. The quality of collected data can also vary significantly, and there may be a need for processing and transformation concerning its quality and consistency. Recording deficiencies often result in missing data sets, which ultimately affects any analysis model (Burmester et al., 2018).

The challenges of using digital assistants can be explained as follows (CloudFlight, n.d.):

- *Approval*: Bots face challenges in gaining acceptance from users, often due to a sense of not feeling valued or past negative experiences, which can make individuals less open to using them.
- *Determination*: Digital assistants lack the ability to assess what is good or bad and cannot evaluate different arguments to make informed decisions. They can only provide responses based on their training.
- *Empathy*: Bots lack the capacity to empathize with customers, as they cannot recognize emotions like anger or sadness. They provide responses unemotionally and are unable to adapt their answers if a user doesn't understand.
- *Situation*: Many assistance systems struggle to maintain context from previous interactions, often starting anew after each question is answered, as they do not remember past interactions.

## The Future of Digital Assistant Technology in the Aviation Industry

Aviation digital assistants are designed with the aim of leveraging the power of artificial intelligence and automation technologies to enhance flight safety, optimize pilots' decision-making processes, improve air traffic management, and foster increased human-machine collaboration.

A range of projects is addressing various scenarios and application areas for the implementation of digital assistants in aviation. These projects are poised to open the door to a significant transformation in the future of the aviation sector. Some of these projects are elaborated upon below.

1. **The HAIKU Project - *Human AI teaming Knowledge and Understanding for aviation safety* (2022-2025);** The HAIKU project has the primary goal of producing and analyzing fully developed prototypes of AI Digital Assistants designed for the aviation sector. The objective is to extract comprehensive human factors design guidelines and methodologies for the creation of safe, efficient, and reliable Digital Assistants for aviation. Furthermore, HAIKU will introduce innovative safety and validation assurance procedures specific to Digital Assistants, with the intention of streamlining their early integration into aviation systems in cooperation with aviation stakeholders and regulatory authorities. According to HAIKU official website, HAIKU will create Digital Assistants across six distinct use cases (Haiku Project, n.d.):
    - *Flight Deck Startle Response:* A Digital Assistant for Urban Air Mobility (UAM) is designed to support UAM coordinators by handling repetitive tasks, monitoring traffic, and addressing non-standard situations during planning and operations. This specific use case is led by LiU, with collaboration and support from LFV, and leverages existing technology developed at LiU, including UTM City and SOMA-AI. A Digital Assistant situated on the flight deck is intended to aid pilots in responding to 'startle response' adverse events.
    - *Flight Deck Route Planning/Replanning*: Another Digital Assistant on the flight deck is aimed at assisting pilots in route planning and re-planning.
    - *Digital and Remote Towers:* Within tower operations, including remote tower setups, a Digital Assistant is envisioned to support controllers in managing routine and repetitive tasks related to aircraft on approach.
    - *Urban Air Mobility:* A Digital Assistant for Urban Air Mobility (UAM) serves the purpose of aiding UAM coordinators in handling repetitive tasks, monitoring traffic, and addressing non-standard situations during planning and operations. LiU takes the lead in this use case, with backing from LFV, and it capitalizes on existing technology developed at LiU, such as UTM City and SOMA-AI.

   - *Airport Safety Watch*: In airport environments, a Digital Assistant is intended to assist in safety management by aiding in data collection, categorization, analysis, and visualization.
   - *Airport Spreding Virus Prevention:* Additionally, at airports, a Digital Assistant is designed to assist in monitoring conditions related to risk factors, such as infectious diseases (e.g., COVID-19), and in preventing these by providing passengers with information on safe paths.
2. **SafeTeam (2022-2025);** SafeTeam is actively advancing research in the realm of human factors concerning the utilization of digital assistants in aviation. This research entails gaining a more profound understanding of the technology and procedures that will support the integration of AI tools into aviation operations, ultimately enhancing human cognitive abilities and the potential for automation. Additionally, the project will delve into the issues of approval and certification, particularly focusing on human proficiency in operating advanced AI tools and the transparency of AI operations. Some objectives of this project are as follows (SafeTeam, n.d.):
   - Advocate a human-centered approach to incorporate automation into various air traffic operations,
   - Suggest methodologies for evaluating and overseeing system performance, with a particular emphasis on safety and resilience, to facilitate effective collaboration between humans and machines,
   - Advance the development of Digital Assistants for aviation operations to enhance human performance throughout all stages of development, testing, validation, and verification,
   - Contribute to the establishment of regulatory and certification requirements for automation tools, addressing both market demands and societal acceptance.
3. **FlytX (2019-2025):** Thales' research and development division is working on a FlytX avionics suite for future business jet cockpits, featuring a touchscreen interface and an artificial intelligence virtual pilot assistant. Initially designed for helicopters, the FlytX system is set to be certified and deployed for the French military's Airbus H160Ms in 2022. Thales is now adapting this technology for business jets, aiming to compete with established players like Honeywell and Collins. Thales' approach integrates multiple avionics components into FlytX, reducing the need for extra hardware and dedicated controls, ultimately optimizing size, weight, and power (SWAP).

This integration and virtualization approach results in a significant SWAP reduction and reduces reliance on line-replaceable units. FlytX maintains a flexible and crew-centered architecture, with data integrated into a digital cloud for specific aircraft types and flight routes, available to pilots on a per-flight basis. The integration of artificial intelligence aims to introduce a virtual assistant to assist pilots in routine tasks. For instance, when receiving voice commands from air traffic controllers like "turn left heading 2-7-0," the assistant interprets and suggests the necessary flight control adjustments on the display, allowing the pilot to easily accept or reject them (Avionics International, 2019).

4. **HARVIS - Human Aircraft Roadmap for Virtual Intelligent System (2019-2035):** The HARVIS (Human Aircraft Roadmap for Virtual Intelligent System) project will be able to help support pilot decision-making in complex situations using cognitive computing algorithms. The HARVIS project will also present a concept paper on the future of artificial intelligence in the cockpit, including a series of realistic scenarios where a digital assistant could assist flight operations. This digital assistant is expected to replicate human-like qualities, such as adaptability, interactivity, and contextual understanding, to enhance decision-making processes (Duchevet et al., 2020; Harvis, n.d.).
5. **Honeywell (OperA & DARWIN):** Honeywell will set to lead two consortiums: Project *OperA* (*Operate Anywhere*), which aims to integrate advanced air mobility (AAM) into European Air Traffic Management (ATM) and U-space, focusing on developing autonomy-enabling technologies for AAM aircraft; and the second project, *DARWIN (Digital Assistants for Reducing Workload & Increasing collaboration)*, which explores AI-based automation for cockpit and flight operations. Within DARWIN, Honeywell seeks to create AI-powered digital assistants and a human-AI collaboration framework to support extended minimum crew operations (eMCO), single pilot operations (SPO), and eventually fully autonomous flights (Honeywell Aerospace, 2023).

## CONCLUSION

It is noteworthy that artificial intelligence (AI) is poised to play a significant role in the future technological landscape of aviation personnel. AI's future in aviation holds exciting possibilities, with advancements in machine learning and data analytics promising more sophisticated applications in areas like manufacturing and surveillance. At present, current artificial intelligence (AI) in the form of Machine Learning (ML) primarily provides information, which can be treated as an artifact or contributes to the functioning of an artifact. In practical terms, it often amounts to an additional layer of automation. However, looking forward to the future, especially in the 2030-2040 timeframe, the anticipated emergence of digital assistants (DAs) promises to extend beyond these capabilities. Present AI support systems do not engage in dialogues, but future DAs are expected to initiate conversations, thereby evolving AI from a mere information provider to a collaborative crew member, and, in some instances, even a crew supervisor. This transformation implies that future digital assistants must not only comprehend the task at hand and its contextual factors, but also possess a certain level of understanding of human culture (Kirwan, 2023).

In aviation, digital assistants are designed to enhance flight safety, optimize pilot decision-making processes, improve air traffic management, and provide a more personalized experience. For example, they can enhance flight safety by instantly providing pilots with information such as weather conditions and fuel levels. Additionally, digital assistants have the potential to reduce the workload of aviation personnel and enhance operational efficiency. This, in turn, can assist airlines and air traffic controllers in making better decisions. The use of digital assistants in aviation holds significant potential. However, for this technology to be successfully implemented, digital assistants must be developed with the capability to understand human culture, interact with humans, and perform complex tasks.

In conclusion, the development of artificial intelligence in aviation enters our lives not only as a competitive advantage but also as a groundbreaking revolution that shapes the future of the aviation world. As long as the development of artificial intelligence-supported virtual assistants continues, it will be inevitable for large-scale applications for flight experiences in the aviation field to be included in our lives. So fasten your seat belts and get ready for virtual assistant-supported travel experiences.

## REFERENCES

Abubakar, M., EriOluwa, O., Teyei, M., & Al-Turjman, F. (2022). AI Application in the Aviation Sector. In *2022 International Conference on Artificial Intelligence of Things and Crowdsensing (AIoTCs)*. IEEE.

Airbus Aircraft. (2021). *Airbus launches Mission+, the first Electronic Flight Assistant that maximises flight efficiency*. Airbus Aircraft. https://aircraft.airbus.com/en/newsroom/press-releases/2021-11-navblue-an-airbus-company-launches-mission

Airports Council International. (2023). *ACI World confirms top 20 busiest airports worldwide*. Airports Council International. https://aci.aero/2023/07/19/aci-world-confirms-top-20-busiest-airports-worldwide/

Avionics International. (2019). *FlytX, Virtual AI Assistant in Development for Future Business Jet Cockpits*. Avionics International. https://www.aviationtoday.com/2019/12/06/flytx-virtual-ai-assistant-development-future-business-jet-cockpits/

BEA. (2012). *Air France 447 Final Report*. BEA. https://bea.aero/docspa/2009/f-cp090601.en/pdf/f-cp090601.en.pdf

BEA. (2022). *Aircraft Accident Investigation Report B737- MAX 8, ET-AVJ*. BEA. https://bea.aero/fileadmin/user_upload/ET_302__B7378MAX_ACCIDENT_FINAL_REPORT.pdf

Burmester, G., Ma, H., Steinmetz, D., & Hartmannn, S. (2018). Big data and data analytics in aviation. *Advances in Aeronautical Informatics: Technologies Towards Flight*, *4*(0), 55–65. doi:10.1007/978-3-319-75058-3_5

Cam, A. C., & Durmaz, V. (2018). Dijital Havacılık: Güncel Uygulamalarla Gelecekteki Yolcu Deneyimleri. *Electronic Turkish Studies*, *13*(26), 251–266. doi:10.7827/TurkishStudies.14163

Chang, S., Wang, Z., Wang, Y., Tang, J., & Jiang, X. (2019). Enabling technologies and platforms to aid digitalization of commercial aviation support, maintenance and health management. In *2019 International Conference on Quality, Reliability, Risk, Maintenance, and Safety Engineering (QR2MSE)*. IEEE. 10.1109/QR2MSE46217.2019.9021222

CloudFlight. (n.d.). The limits and opportunities of digital communication assistants. *Cloud Flight.* https://www.cloudflight.io/en/blog/digital-communication-assistants-capabilities/

Cook, A., & Rivas, D. (2016). *Complexity science in air traffic management.* Routledge. https://books.google.com.tr/books?hl=tr&lr=&id=kcRJDAAAQBAJ&oi=fnd&pg=PP1&dq=cook+rivas+2016&ots=t67Hr2dSZI&sig=FKRDYROMHz7Tc82LPHnVF-rmyAA&redir_esc=y#v=onepage&q=cook%20rivas%202016&f=false

Çukur, F. T. (2020). Sivil Havacılıkta Büyük Veri Teknolojisi ve Uygulamaları. *ThinkTech Araştırma Raporu Ocak 2020.* https://thinktech.stm.com.tr/uploads/docs/1608904343_stm-sivil-havacilikta-buyuk-veri.pdf

DataBeacon. (n.d.). *Victor5 A Digital Assistant Platform for Air Traffic Control Operations.* Databeacon. https://databeacon.aero/

Digital Adoption. (2023). Airline digital transformation: A digital-centric transition. *Digital Adoption.* https://www.digital-adoption.com/airline-digital-transformation/

Haiku Project. (n.d.). HAIKU. *Human AI teaming Knowledge and Understanding for aviation.* https://haikuproject.eu/

Harrison, A., Popovic, V., & Kraal, B. (2014). The future passenger experience: A shift from physical to virtual design. In *Proceedings of the International Conference on Kansei Engineering and Emotion Research (KEER) 2014.* Eindhoven University of Technology.

Hauswald, J., Laurenzano, M. A., Zhang, Y., Li, C., Rovinski, A., Khurana, A., Dreslinski, R. G., Mudge, T., Petrucci, V., Tang, L., & Mars, J. (2015). Sirius: An Open End-to-End Voice and Vision Personal Assistant and Its Implications for Future Warehouse Scale Computers. *SIGPLAN Notices, 50*(4), 223–238. doi:10.1145/2775054.2694347

Honeywell Aerospace. (2023). Honeywell To Help Digitize The Skies Through New Research And Development Programs In Europe. *Honeywell Aerospace.* https://aerospace.honeywell.com/us/en/about-us/press-release/2023/06/honeywell-to-help-digitize-the-skies-through-new-research-and-development-programs-in-europe

IGA Sustainability Report. (2019). *A New Airport Experience*. Istanbul Airport IGA. https://www.igairport.aero/media/osapjfn3/surdurulebilirlik2019-en.pdf

Innova. (2023). Havacılık sektöründe dijital dönüşüm trendleri. *Innova*. https://www.innova.com.tr/blog/Havacilik-sektorunde-dijital-donusum-trendleri

IVisa. (2023). What is IATA Travel Pass? *IVisa Featured on CNN*. https://www.ivisa.com/blog/what-is-iata-travel-pass

Jadoon, W. (2023). The rise of AI in aviation: Trends shaping the future. *Vantage Plus*. https://www.vantageplusanalytics.com/the-rise-of-ai-in-aviation-trends-shaping-the-future

Jaffer, S., & Timbrell, G. (2014). Digital strategy in airports. In *Proceedings of the 25th Australasian Conference on Information Systems*. ACIS/Auckland University of Technology.

Kirwan, B. (2023). The Future Impact of Digital Assistants on Aviation Safety Culture. *Human Interaction and Emerging Technologies (IHIET-AI 2023). Artificial Intelligence and Future Applications*, *70*(70), 77–87. doi:10.54941/ahfe1002932

Krakovsky, M. (2018). Artificial (emotional) intelligence. *Communications of the ACM*, *61*(4), 18–19. doi:10.1145/3185521

MITRE. (n.d.). *Digital Copilot*. MITRE. https://www.mitre.org/our-impact/intellectual-property/digital-copilot

Molchanova, K. (2020). A Review of Digital Technologies in Aviation Industry. *Logistics and Transport*, *47*(3-4), 69–77.

Munich Airport. (2023). *First Smart Baggage Trolleys introduced at Munich Airport*. Munich Airport. https://www.munich-airport.com/press-making-shopping-more-digital-19856087

NavBlue. (n.d.). Ready to enter the era of Electronic Flight Assistant? *NavBlue*. https://www.navblue.aero/product/mission/

Olcay, S. (2019). *Sosyal İhtiyaçların Karşılanmasında Yeni Medya Uygulama Örneği Olarak Dijital Asistanlar* [Master's Thesis, Istanbul Aydın University]. YökTez Database.

Pegasus Airlines. (n.d.). *Pegasus'un Sanal Asistanı FlyBot Hizmetinizde*. Pegasus Airlines. https://www.flypgs.com/pegasusun-sanal-asistani-flybot-hizmetinizde

Perez, D. (2021). AI Digital Assistants for Air Traffic Controllers will revolutionize operational rooms. *DataScience Aero Blog*. https://datascience.aero/ai-digital-assistants-air-traffic-controllers-operational-rooms/

Pieraccini, R. (2021). *AI Assistants*. MIT Press. https://mitpress.mit.edu/9780262542555/ai-assistants/

Precedence Research. (2022). *Artificial Intelligence in Aviation Market Global Industry Analysis, Size, Share, Growth, Trends, Regional Outlook, and Forecast 2022 – 2030*. Precedence Research. https://www.precedenceresearch.com/artificial-intelligence-in-aviation-market

Ramos, D. (2018). Voice Assistants: How Artificial Intelligence Assistants Are Changing Our Lives Every Day. *Smartsheet*. https://www.smartsheet.com/voice-assistants-artificial-intelligence

SafeTeam. (n.d.). About Safe Team. *SafeTeam*. https://safeteamproject.eu/

Sanal Savunma. (2021). Havayollarının elindeki gizli hazine; Big Data (Büyük Veri). *Sanal Savunma*. https://www.sanalsavunma.com/havayollarinin-big-data-buyuk-veri/

Silling, U. (2019). Aviation of the future: What needs to change to get aviation fit for the twenty-first century. In *Aviation and Its Management-Global Challenges and Opportunities*. IntechOpen. https://www.intechopen.com/chapters/65435. doi:10.5772/intechopen.81660

SmartRep. (2023). *"Philos": Athens International Airport's new digital assistant*. SmartRep. https://smartrep.gr/case_study/philos-athens-international-airports-new-digital-assistant/

Tensator. (2023). *Tensator Virtual Assistant*. Tensator. https://www.tensator.com/solutions/tensator-virtual-assistant/

The Moodie Davitt Report. (2019). Frankfurt Airport to pilot 'robot concierge' in digitalisation drive. *The Moodie Davitt Report*. https://www.moodiedavittreport.com/frankfurt-airport-to-pilot-robot-concierge-in-digitalisation-drive/?format=pdf

Thompson, R. (2018). *Three ways AI can improve Aviation Security*. LinkedIn. https://www.linkedin.com/pulse/three-ways-ai-can-improve-aviation-security-richard-thompson/

Turkish Airlines Terminal. (n.d.). *Seyahat Asistanı 'Boti'*. Turkish Airlines Terminal. https://terminal.turkishairlines.com/portfolio_page/seyahat-asistani-boti/

UTED. (2019). Dijital Dönüşüm ve Geleceğin Havalimanları. *Uçak Teknisyenleri Derneği*. https://www.uted.org/d%C4%B1j%C4%B1tal-d%C3%B6n%C3%BC%C5%9F%C3%BCm-ve-gelece%C4%9F%C4%B1n-haval%C4%B1manlar%C4%B1/

WithZoey.com. (2023). We make empathy digital. *Zoey*. https://withzoey.com/

## ENDNOTES

1 Deep Learning: Deep learning, which falls under the umbrella of machine learning, is founded upon meticulously designed neural networks. This methodology encompasses the organization of neural networks into consecutive layers, facilitating a continual process of data assimilation, especially when the objective is to construct models from unstructured data. Deep learning leverages sophisticated neural networks specifically engineered to emulate the intricate operations of the human brain, rendering it adaptable for the resolution of ambiguously defined abstractions and complex problems without infringing upon academic integrity.

2 Zoey: Passenger-oriented digital assistant (WithZoey.com, 2023).

3 The Air France 447 crash occurred in 2009 due to ice crystals, rendering the aircraft's airspeed sensors ineffective and disconnecting the autopilot (BEA, 2012).

4 These two accidents were caused by design flaws and specifically the use of flight control software known as MCAS (BEA, 2022).

# Chapter 2
# Emerging Role of Artificial Intelligence (AI) in Aviation:
## Using Predictive Maintenance for Operational Efficiency

**Tereza Raquel Merlo**
https://orcid.org/0000-0002-2042-5415
*University of North Texas, USA*

## ABSTRACT

*This book chapter offers an examination of the transformative influence of Artificial Intelligence (AI) within the aviation sector, focusing specifically on its application in predictive maintenance for enhancing operational efficiency. Through the investigation of current research and industry trends, and the utilization of a literature review as its methodological framework, this chapter elucidates the transformative impact of AI-driven predictive maintenance strategies on aviation operations. It explores how AI algorithms analyze vast amounts of data to predict potential equipment failures, enabling proactive maintenance interventions that minimize downtime and optimize fleet performance, presenting an analysis of the implications of AI integration in aviation. Notably, the integration of AI-driven technologies in critical areas such as flight planning, predictive maintenance, and air traffic management is highlighted, showcasing the significant advancements that have reshaped the aviation landscape. Furthermore, the chapter adopts a user-centric perspective, offering a critical assessment of the challenges and considerations inherent in AI implementation in aviation. This includes examining issues pertaining to human-machine interaction, trust in AI systems, and the evolving dynamics of job roles within the industry. Overall, the chapter provides a comprehensive overview of AI's impact on aviation, offering valuable insights for aviation professionals, policymakers, and researchers seeking a deeper understanding of the profound changes driven by AI within the aviation domain.*

DOI: 10.4018/979-8-3693-0732-8.ch002

## INTRODUCTION

As the aviation industry continues to embrace digital transformation, the integration of artificial intelligence (AI) and data analytics has emerged as a game-changer. The rapid advancement of digital technologies has brought about significant transformations across various industries, and the air transportation sector is no exception. The technological revolution introduced in the aviation industry by AI redefines operational paradigms and efficiency benchmarks. Among the myriad applications of AI in aviation, predictive maintenance emerges as a pivotal domain, offering a proactive approach to preserving aircraft integrity and optimizing operations and decision-making. This chapter delves into the crucial domain of air traffic management (ATM) and explores how digital innovation is revolutionizing the way airspace is managed, flights are coordinated, and safety is ensured.

The integration of Artificial Intelligence (AI) in the aviation industry has introduced transformative changes, significantly affecting various stakeholders and the overall user experience. Benbya, Davenport, and Pachidi (2020) provides a comprehensive overview of the state of Artificial Intelligence (AI) in an organizational context, analyzing the key aspects of accessibility, technological impact, real-world experiences, and a brief list of challenges and opportunities, highlighting the ethical concerns and the potential of AI in enhancing decision-making for business transformation. Understanding the profound implications of AI implementation necessitates a comprehensive examination of existing research and studies.

This chapter embarks on a journey to explore the emerging role of AI in aviation, with a keen focus on its application in predictive maintenance. Through a nuanced examination of current practices, technological advancements, and industry insights, it aims to unravel the transformative potential of AI-driven predictive maintenance strategies. By delving into the intricacies of data analytics, machine learning algorithms, and real-time monitoring systems, the goal is to contribute to uncover how AI facilitates predictive insights that enable timely maintenance interventions, ultimately enhancing operational efficiency and aircraft reliability.

In the following pages, we will explore case studies, industry best practices, and technological innovations that showcase the tangible benefits of integrating AI into maintenance workflows. Moreover, discussing challenges and opportunities inherent in the adoption of AI-driven predictive maintenance, considering factors such as data quality, regulatory compliance, and workforce readiness. The exploration of existing research and studies within the chapter serves to illuminate AI's role in enhancing key aspects of aviation, including safety, operational efficiency, the integration of AI-powered predictive maintenance systems into existing aviation maintenance practices, benefits of cost reduction, improved safety, and enhanced aircraft reliability, and the decision-making processes. This chapter aims to provide aviation professionals, policymakers, and researchers with a comprehensive understanding of the transformative potential of AI in predictive maintenance. By illuminating the synergies between AI technologies and aviation maintenance practices, this chapter aims to inspire further innovation, foster collaboration, and drive forward the pursuit of operational excellence in the aviation industry.

## METHODOLOGICAL FRAMEWORK

The literature review serves as an invaluable methodological approach in understanding AI's impact on aviation and the user experience. By synthesizing peer-reviewed studies, this section sheds light on the transformative potential of AI in aviation, the importance of human-AI interaction, and the evolving roles and expertise of aviation professionals. Emphasizing user perspectives and needs ensures that AI integration enhances aviation safety, efficiency, and user satisfaction, ultimately guiding the industry towards a harmonious human-AI collaboration. The literature review explores various academic and industry publications, covering topics such as AI-driven predictive maintenance, intelligent flight operations, passenger experience enhancement, and air traffic management optimization. By examining these sources, we gain a comprehensive understanding of the current landscape of AI and data analytics in aviation and the impact on user experiences

Numerous studies have explored AI-driven technologies that revolutionize aviation operations. Notably, AI's role in flight planning and optimization (Wilson et al., 2019), predictive maintenance for enhanced aircraft reliability (Liu et al., 2020), and AI-based air traffic management systems (Bandyopadhyay et al., 2018) stand out as pivotal advancements in aviation efficiency and safety. AI's introduction in aviation raises questions about the shifting roles of aviation professionals. User studies conducted by Smith et al. (2019) reveal the changing skill sets and expertise required to effectively collaborate with AI systems. Ensuring user empowerment and training (Patterson et al., 2022) are pivotal in maximizing AI's potential while preserving the importance of human knowledge and experience in critical aviation tasks.

The work of Samatas, Moumgiakmas, and Papakostas (2021), focusing on predictive maintenance-bridging artificial intelligence and Internet of Things (IoT) in aviation, discusses the automation process and the digital evolution imposing radical changes in the way processes are being automatized. The authors examine big data and data processing using machine learning (ML) for high quality algorithms and prediction. Modern techniques applying PdM (Predictive Maintenance) and advanced sensors are examples of ML applications in aviation maintenance becoming more commonly used for efficiency. It is argued that:

*machine learning has evolved into a powerful tool that implements high quality intelligent algorithms for prediction. A feature of ML is the ability to process big data and therefore effectively find hidden correlations between them. These data can be complex and produced in constantly changing environments. Such environments are heavy industry, the automotive industry, aviation/aeronautics, business assets and even various household appliances. (p. 414)*

There are reportedly numerous predictive maintenance strategies used across aviation industries, from reactive to proactive seeking to respond to failures or preventing them from occurring. PdM is applied with many levels of efficiency and is normally costly. It can "exploit networks of sensors to gather data that can be analyzed to identify the health and degradation of a given system." (Staton, Munir, Ikram, and El-Bakry, 2023, p. 217), enabling a system analysis that applies pattern recognition and statistical analysis responsible for a predictive reliable approach in aircraft maintenance strategy.

It is understood that the user perspective on AI is particularly relevant in Unmanned Aircraft Systems (UAS) operations. Gaining insights into the user experience of drone pilots and controllers (Buss et al., 2020) elucidates challenges, such as situational awareness and workload management, and informs the development of user-centered

AI solutions in this domain. Effective human-AI interaction is critical for seamless integration and user acceptance. Research by Lee et al. (2018) highlights the significance of designing AI systems that enhance user trust, transparency, and controllability, thereby fostering positive user experiences. Identifying user preferences and expectations (Liang et al., 2021) plays a crucial role in shaping AI's deployment in aviation, ensuring the alignment of AI capabilities with user needs.

## AI Driving Strategic Decisions in Aviation

Ranganathan (2023) presents an argument that exemplifies the use of AI in decision-driven initiatives with the goal to optimize profitability through a personalized C-Suite program. The author discusses the Personalization AI System (PAI) which, combined with Revenue Management (RM), creates the multi-dimensional data structure, called micro-segments, that enables a personalized, on-demand, system promoting short-, medium-, and long-term value tradeoffs. Ranganathan (2023) claims that: "The PAI system based on microsegments is a great tool for guiding effective Personalization while maintaining the actions needed for these tradeoffs. Within the context of the microsegments, the system identifies all the KPIs that govern the business—transactions volumes, yield, RASK, revenues, costs, profits, market share, capacity share, CLV and so on. (p. 169).

Kulida and Lebedev (2020) presented a report discussing the use of artificial intelligence in aviation with the purpose of tracing the current and potential trends and technologies that can be used in problem solving for civil and military aviation. The authors researched domestic and foreign publications identifying several areas for the use of AI in the aviation sector, including intelligent crew interface, support for operational decisions by the crew, collection, processing, and analysis of air traffic information, optimization of the airspace structure and the flow of aircraft, optimization of flight routes, development of means for detecting and preventing collisions between aircrafts and unnamed aerial systems; aviation computer training complexes, diagnostics or airborne components and assemblies, automation and autonomy of management, and solving combat missions. The conclusion of Kulida and Lebedev's study raises questions about the intricate dynamic of how AI systems are used in the decision-making process and all the possible contingencies related to the operational aspects of decision-making involving humans, which inevitably evidenced neural networks vulnerabilities and serious errors.

According to Kulida and Lebedev (2020), tests are being conducted by the Defense Developmental Agency (DARPA) utilizing neural network algorithms that could lead to the development of an AI system able to conduct maneuverable air combat. The authors exemplify the importance and use of the system citing the case of Boeing Corporation and its planned test on autonomous passenger aircraft, the Airbus Corporation, as the beginning of a single-pilot aircraft where the second pilot is replaced by AI, which could evolve to the next phase of a fully autonomous aircraft.

Cheng and Li (2016) reviewed the classic models and implementation methods of Artificial Neural Network (ANN) and analyzed the key issues of implementation based on Field Programmable Gate Array (FPGA) amid intelligent information processing system in aviation field. The study emphasizes the development of an artificial neural network algorithm and its evident superior performance. This research defends that the current stage is only the beginning phase of a revolutionary process that the performance of the neural networks relies on computer cluster. The authors conclude by asserting the relevance in the concepts and implementation of ANN and FPGA in the aeronautical field, declaring that: "If we can solve those problems, airborne intelligent information processing system will no longer just an idea, our aircraft will become more intelligent and more secure." (p. 410).

Digital platforms are facilitating collaborative decision-making processes involving airlines, airports, air traffic controllers, and other stakeholders. There are numerous examples of how such collaboration optimizes resources, minimizes delays, and enhances the overall passenger experience, facilitated by the integration of information systems (Kumar, 2022).

Some critical issues related to cybersecurity in digitalized ATM systems involve potential systems vulnerabilities, and the strategies in place to safeguard against cyber threats. Ensuring the resilience of air traffic management operations are critical; while existing regulatory challenges posed by the rapid adoption of digital innovation in air transportation should remain focused on the investigation into how aviation authorities are adapting regulations to ensure safety and security while fostering innovation, security and reliability are critical priorities. An additional aspect to be contemplated includes human-machine collaboration when exploring the role of air traffic controllers' pilots in a digitally transformed landscape. The need for an optimal human-machine collaboration is unquestionable, where automation assists professionals in decision-making processes.

## AI-Driven Predictive Maintenance systems and Machine Learning Applications

AI-driven predictive maintenance systems are transforming aircraft maintenance practices in unmeasurable ways. These systems use sensors and data analytics to monitor the condition of aircraft components in real-time. When abnormalities are detected, maintenance crews can be alerted, allowing for timely repairs and minimizing the risk of in-flight failures. The other aspect being favored by the advances of AI in aviation is related to materials science. AI and materials science intersect in the aviation industry helping identify and develop advanced materials that are lightweight, durable, and well-suited for aviation applications.

Aviation industry generates massive amounts of data that could potentially be used to benefit the industry, the company, and customers, if effectively managed. However, research has shown that most data generated are poorly managed or not organized and used at all. An example, Harvard Business Review brings an article by James Wilson et al. illustrating how companies are using machine learning to improve their efficiency. Wilson, Sachdev, and Alter (2016) stated that:

*Machine learning can also aid human decision making by discovering meaningful patterns in factory data that people would otherwise be unable to find. Consider Sight Machine again: By analyzing data for patterns that occur before trouble hits, the company's systems help manufacturing engineers anticipate and prevent problems. For one client deploying a new robotic manufacturing line, Sight Machine was able to reduce downtime by 50% and increase performance by 25% within one month — far better than the 1%–2% performance increases typical of the client's industry. (para. 11).*

AI is being used and will continue to be used to optimize air traffic flows, reduce delays, and enhance safety through systems like the Collaborative Air Traffic Management Technologies (CATMT), effectively employing machine learning to analyze vast amounts of aviation data to identify trends, predict equipment maintenance needs, and optimize airspace utilization. Data analytics in aviation also helps with weather forecasting. AI algorithms are used to improve weather forecasting, helping pilots and air traffic controllers make informed decisions to avoid adverse weather conditions. Additional benefits include the implementation of systems, like NextGen, which aims to reduce flight delays, save fuel, minimize environmental impact, and enhance overall aviation safety and efficiency. The advances in avionics help pilots in navigating complex flight paths and responding to emergency situations. Additionally, AI's role improves navigation and collision avoidance systems, reducing the risk of accidents and enhancing passenger safety, impacting avionics systems, and guiding advancements towards autopilots powered by AI.

## AI Application as a Critical Factor Impacting the Aviation Industry: Real Case Applications

A better investigation and understanding of the factors involved in the transformative role of artificial intelligence in the design, development, and management of aircraft and aviation systems will enable paths to explore how AI-powered simulations, computational models, and data analytics have revolutionized the aviation industry's approach to crafting safer, more efficient, and environmentally friendly aircraft. Merlo (2024) concurs that:

*AI has surpassed mere imitation; it has become a catalyst for innovation, efficiency, and strategic advantage [...]. From machine learning and neural networks to natural language processing and computer vision, the building blocks of AI are assembled to create intelligent systems that can analyze, learn, and adapt. The power of AI is not confined to its technical prowess alone; it resides in its symbiotic relationship with the ever-evolving technological landscape. The interconnected web of technologies, such as cloud computing, big data analytics, and the Internet of Things (IoT) intertwines with AI, amplifying its capabilities and expanding its potential applications. (p. 3)*

The enhancements in AI aerodynamics, for example, allows for a better understanding of how AI algorithms are utilized to optimize the shape and design of aircraft wings, fuselage, and other components. AI simulations and computational fluid dynamics models assist in fine-tuning aerodynamics, leading to reduced drag, improved fuel efficiency, and enhanced flight performance, not to mention real-world examples of aircraft models that have benefited from AI-driven aerodynamic improvements. Delta airlines exemplifies how an aviation business has been using AI-driven machine learning strategically.

## The Delta Airlines: Leveraging AI and the Approach to Predictive Maintenance

According to Eric Snell, Delta's Senior Vice-President – Operations & Customer Center (OCC):

*The proprietary AI-driven platform analyzes millions of operational data points - from aircraft positions to flight crew restrictions to airport conditions - to create hypothetical outcomes that help Delta's professionals make critical decisions before, during and after large-scale disruptions, like severe winter weather or a volcanic eruption. It is also a post-mortem tool that can be used to identify how better decisions could have been deployed in a given situation. As more real data is integrated, the innovative platform architecture applies machine learning to arm decision makers with the best range of options to ensure safe, reliable travel. (Nothing artificial, 2020, para. 3).*

Delta's investment and prioritization of AI for over a decade has allowed the company the use of predictive analytics to filter data and provide actionable insights into situations, including maintenance, personnel, technologies, and technical issues. In March of 2018, this approach to relying on predictive advanced data and analytics systems prevented a disastrous situation that could have resulted in the delay of nearly 1,200 aircrafts and flights cancellations. (Nothing artificial, 2020).

Delta Air Lines is one of the largest airlines in the world, and despite its commitment to superior services and customer interfaces, faced challenges in maintaining its extensive fleet of aircraft efficiently. Costs with maintenance were reportedly prompting unexpected equipment failures, leading to costly delays and safety concerns. The adoption of AI-driven predictive maintenance resolved those issues. Delta partnered with an AI technology company to implement a predictive maintenance system that would address three main aspects: **a)** Data Collection – Adopting sensors and IoT devices that, once installed on Delta's aircraft, allow the airlines to collect vast amounts of real-time data, including engine performance, temperature, pressure, and component wear; **b)** Data Analytics – AI algorithms were developed to analyze the data continuously. These algorithms could detect anomalies and patterns indicative of equipment degradation or potential failures. Machine learning models were trained to predict when specific components would require maintenance; and **c)** Predictive Alerts – The AI system generated predictive alerts for maintenance crews when it detected issues that needed attention. These alerts were based on the actual condition of aircraft components rather than scheduled maintenance intervals. The objective was to benefit from the use of AI in predictive maintenance to:

- Reduce Downtime – Unplanned maintenance events and delays due to equipment failures were significantly reduced. Aircraft spent more time in the air and less time on the ground for repairs.
- Reduce costs – Delta experienced substantial cost savings by avoiding costly last-minute repairs and minimizing the need for spare parts inventory.
- Improve Safety – The proactive nature of predictive maintenance enhanced flight safety by preventing potential failures before they could compromise the integrity of the aircraft.
- Improve Passenger Experience – Fewer flight delays and cancellations resulted in a better passenger experience, leading to increased customer satisfaction and loyalty.

Executive leaders of Delta Airlines claim that:

*Over the past several years, Delta has led the airline industry in transforming the customer experience by introducing the first end-to-end biometric terminal in the US at Atlanta's international airport, real-time RFID bag tracking and automatic check-in via the Fly Delta mobile app, an alliance that will empower customers with seamless in-cabin connectivity experience, more efficient and high-tech automated screening lanes, and a groundbreaking app that helps Delta pilots avoid turbulence for a more comfortable flight." (Nothing..., 2020, para. 9).*

The case of Delta airlines demonstrates how AI-driven predictive maintenance can have a transformative impact on an airline's operations, leading to cost savings, enhanced safety, and improved customer service. It serves as a compelling example of AI's practical application in the aviation industry. Kashyap (2019). Delta's decision to focus on to the use of AI in the predictive maintenance strategy prompted the expansion of the program to include areas for flight optimization, fuel efficiency, and staff scheduling, as examples. Wilson et al (2016) advocates that AI introduces valuable insights into reducing and/or preventing disruptions and flights delays, improving customer experience, and increasing the overall flight experience. Wilson defends that: "with machines as sidekicks, though, people can more quickly find valuable insights buried in big data," (para. 3). Ranganathan (2023) presented a study for driving personalization systems in airlines services, defending that personalization programs based on AI systems helps the C-Suite, the personalization of customers services, improves the organizational ability to assess risks and opportunities, and drives superior revenue. The author advocates for the use of an AI system that is grounded on Big Data to support decision-making: from forecasting to actionable recommendations using personalized strategies for Revenue Management (RM). The RM driven personalization in microsegments designed by the author aggregates insights from an operational level in the business translating in millions of dollars of incremental revenue.

## Air Traffic Management using ML and AI: the case of the Israeli Defense Force (IDF) and the EASA

Evolution of Air Traffic Management: Air traffic management systems have been facing a transformative process in the face of challenges faced by traditional systems in coping with the growing demands of air travel, the increase air traffic, complex technological changes, and the demand to accommodate a large flux of people. Digital Technologies Reshaping ATM, the cutting-edge digital technologies that are reshaping air traffic management, such as artificial intelligence, machine learning, blockchain, and the Internet of Things (IoT) are applied to enhance safety, efficiency, and sustainability, and disrupt the traditional way of flying. It is also enabling superior data-driven decision making that emphasizes the emerging aircraft system and focuses on the role of data in modern

ATM systems and the importance of real-time data collection, analysis, and sharing in improving route planning, weather prediction, and incident response-and time response. The Unmanned Aerial Vehicles (UAVs) Integration, for example, delves into the challenges and opportunities associated with integrating drones and other unmanned aerial vehicles into the existing airspace. It is important to highlight the role of digital innovation in enabling safe and efficient coexistence between traditional aircraft and UAVs.

Although relatively new and having a limited scope, the use of AI in the aviation industry is revolutionary and on a path of fast development in military defense systems. In many countries across the globe the increasing deployment of AI technologies in defense strategies is promising. More recently, the Israeli Defense Force (IDF) claimed to be using advanced AI in their defense in the war against airstrikes by Hamas[1] (Gold, Faqiri, Regan…et al., 2023). According to reports from the IDF, the Israel military is using AI systems in airstrikes and in the creation of a plan of attack, implementing an AI recommendation system "to crunch vast amounts of data for faster" (Saballa, 2023) and better target identification, systems calculations, the use of drones, and the management of human operators.

With the goal to protect the nation from missile threats, military forces utilize advanced AI and machine learning applications in the detection of radar, interception of missiles, that intercept and destruction of rockets in mid-air. Kulida and Lebedev (2020) presented the result of a study tracing the trends in the development of AI for problem solving in military aviation, concluding that existing factors in system vulnerability of neural networks could lead to serious errors and that human factors are a crucial consideration in the operational decision-making process. Security requirements and numerous contingencies were also cited as critical factors in the successful application of AI in aviation.

Aslan and Tolga (2022) investigated ways in which airlines could improve flight performance and flight safety with minimal maintenance-repair costs, highlighting that big data and emerging technologies in AI are increasingly supporting maintenance-repair-overhaul (MRO) and allowing the improved multi-criteria-decision-making (MCDM) method. The authors conclude that predictive and preventive maintenance should utilize software "Alternative-6" with global values and ranking equal to 1 global measure, translating in the most suitable predictive maintenance application out of the 11 alternative software considered in the study.

Aspects evaluated in the investigation contemplated 13 criteria: Data Security (C1), Compatibility with Existing IT Systems (C2), Project Cost (C3), Commissioning Time (C4), Degree of risk (C5), User-friendliness (C6), Service and Support (C7), Flexibility (C8), Functions/features (C9), Reliability (C10), Vendor Reputation (C11), Product Viability (C12), and Technology Preferences (C13).

The European Union Aviation Safety Agency, EASA, responsible for aviation safety, standardization and regulations also performs investigations, monitoring, and protecting of civil aviation within the EU and associated countries. In 2020 EASA released the Artificial Intelligence Roadmaps: a human-centric approach to AI in aviation. In this document the integration of AI and machine learning into aviation safety and related activities is explores. The AI roadmaps encompass areas of aviation that include safety management, air traffic management, aircraft design and certification, data analysis, crew training, maintenance and inspection, security and thereat detection (EASA, 2020).

EASA designed the AI roadmap in aviation with the purpose of establishing a vision in the aviation domain and creating a foundation for stakeholders interacting with developments in aviation in response to the dimensions of AI. According to EASA regarding AI applications:

*AI will not only affect the products and services provided by the industry; it will also trigger the rise of new business models. This will affect most of the domains under the mandate of the Agency. Its core processes (certification, rulemaking, organization approvals, and standardization) will be impacted. This will in turn affect the competency framework of Agency staff. (p. 3).*

EASA (2020) applies a definition of AI in the roadmap that defends AI as: "any technology that appears to emulate the performance of a human." (p. 4). This approach must include aspects of trustworthiness and ethics as well as the AI application model that is centered on data linked to machine learning (ML) and a data algorithm of deep learning (DL) for improved performance.

The taxonomy of AI roadmap focuses on ML techniques that apply DL, ML, and AI applications supported by expert systems and deeper neural networks (NNs) as part of a complex NN architecture and a hybrid AI in the scope of the roadmap, as illustrated Figure 1- Artificial

*Figure 1. Artificial Intelligence Roadmaps A human-centric approach for AI in aviation. EASA Roadmap*

*Source: EASA Roadmap (2020). Adapted from "Artificial Intelligence Roadmaps A human-centric approach for AI in aviation. EASA Roadmap" by European Union Aviation Safety Agency, 2020, URL. http://www.adfingo.com/easa-ai-roadmap-v10.pdf*

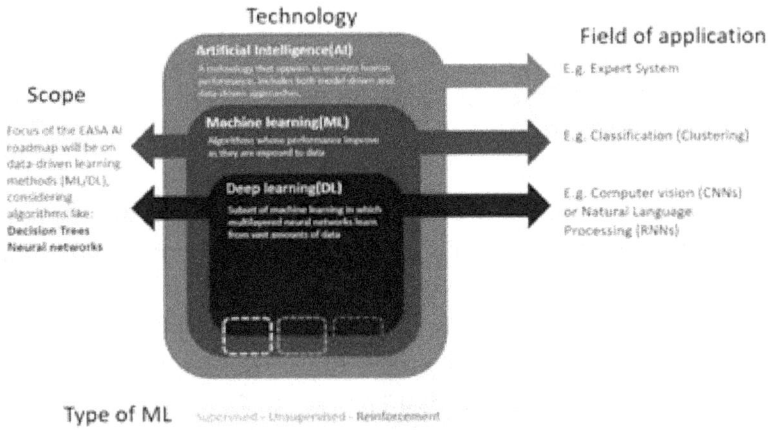


Intelligence Roadmaps A human-centric approach for AI in aviation. EASA Roadmap below:

Some of the challenges identified by EASA in the analysis and design of the AI roadmap contains aspects of competency of staff regarding certification, standards development, rules, etc., including:

- support to the industry, adapting to the regulatory framework and new technologies
- knowledge asymmetry within the industry and certification of AI, among others.
- support to the EU AI strategy and initiatives, seeking to align EU aviation industry standards and regulations with the EU commission supporting research and innovation.

To address those challenges the EASA roadmap identified five top objectives: "1) Develop a human-centric AI Trustworthiness framework; 2) Make EASA a leading certification authority in AI; 3) Support European Aviation leadership in AI; 4) Contribute to an efficient European AI research agenda; 5) Contribute actively to EU AI strategy initiatives." (EASA, 2020, p. 23).

The research analysis of Cheng, Wen, and Li (2016) emphasizes the relevance of an Artificial Neural Network (ANN) in the information processing method in aviation, concluding that traditional airborne computer systems must be combined with neural network processing models resulting in a more intelligent and secure aircraft system.

## CONCLUSIONS AND RECOMMENDATIONS FOR FUTURE WORK

This book chapter contributes to the existing body of knowledge by providing a comprehensive analysis of AI and data analytics in aviation from a user perspective. It bridges the gap between academic research and practical applications, offering valuable insights for researchers, industry professionals, policymakers, and aviation enthusiasts alike. The chapter concludes with recommendations for future research directions and potential areas for further exploration, encouraging ongoing dialogue and collaboration in this rapidly evolving field, particularly in terms of predictive maintenance and the application of advanced technologies.

In addition to the literature review, this study would benefit from the employment of a methodological approach to gather user perspectives. Interviews, surveys, and focus groups are conducted to capture the opinions, needs, and expectations of aviation professionals, passengers, and other relevant stakeholders. By integrating user perspectives, we aim to provide a comprehensive analysis of the benefits, challenges, and future opportunities associated with AI and data analytics in aviation.

By examining these topics, this chapter contributes to the broader narrative of harnessing digital innovation for air transportation and offers valuable insights into the ongoing revolution in the skies. The insights derived from the literature review and user perspectives will enable aviation industry stakeholders to make informed decisions regarding the adoption and implementation of AI and data analytics technologies. By understanding user needs and expectations, organizations can tailor their strategies to optimize operations, improve passenger satisfaction, and drive innovation.

The chapter shed light on the transformative potential of AI and data analytics in aviation from a technologically advanced employment of resources and systems management perspective. It highlights the positive impact of AI-powered systems in enhancing operational efficiency, safety, and ultimately the customer experience, systems management, and increase of profitability. Additionally, it identifies concerns related to security, aircraft training and development, and some of the ethical implications of AI implementation, in addition to highlighting the costs involved in enhanced technological applications and complex (predictive and proactive) data analysis and machine learning solutions.

Future studies should focus on the neural network systems capabilities and challenges with the model of the ANN and the prospects for unnamed aerial vehicles (UAV) and the autonomous aircraft system. To further expand the scope of the study, an analysis centered on users' interfaces and perceptions would favor future investments and initiatives in the use of AI in aviation. Aspects related to safety and the perception of it are important factors to customers and employees as technologies evolve and individuals rely more on neural network systems and algorithms to perform tasks and more companies implement complex systems to enhance performance, increase profitability, and ensure competitiveness. Finally, significant aspects of AI advances in aviation must consider the inexistence of international regulations and the potential consequences of a heavily autonomous aviation system that many professionals and researchers predict as the unavoidable future in the aircraft industry.

## REFERENCES

Aslan, M. E., & Tolga, A. C. (2022). Evaluation of Artificial Intelligence Applications in Aviation Maintenance, Repair and Overhaul Industry via MCDM Methods. In C. Kahraman, A. C. Tolga, S. Cevik Onar, S. Cebi, B. Oztaysi, & I. U. Sari (Eds.), *Intelligent and Fuzzy Systems. INFUS 2022. Lecture Notes in Networks and Systems* (Vol. 504). Springer., doi:10.1007/978-3-031-09173-5_94

Bachtiar, T. A. (2009). *Hamas: Kenapa dibenci Israel?* Hikma.

Bandyopadhyay, T., Sen, S., & Dutta, S. (2018). Artificial Intelligence in Air Traffic Management Systems: A Review. *2018 International Conference on Computing, Power and Communication Technologies (GUCON).*

Benbya, H., Davenport, T. H., & Pachidi, S. (2020). Artificial Intelligence in Organizations: Current State and Future Opportunities (December 3, 2020). [Available at SSRN: https://ssrn.com/abstract=3741983 or http://dx.doi.org/10.2139/ssrn.3741983]. *MIS Quarterly Executive*, *19*(4), 4. Advance online publication.

Buss, A., Pop, C. L., Reid, G., & Kirsch, A. (2020). User Experience in Unmanned Aircraft Systems: A Survey and Research Agenda. *IEEE Access : Practical Innovations, Open Solutions*, *8*, 144421–144434.

Cheng, T., Wen, P., & Li, Y. (2016, December). Research status of artificial neural network and its application assumption in aviation. In 2016 12th International Conference on Computational Intelligence and Security (CIS) (pp. 407-410). IEEE.

European Union Aviation Safety Agency. EASA, (2020). Artificial Intelligence Roadmap: A human-centric approach to AI in Aviation. Retrieved October 12, 2023 from: http://www.adfingo.com/easa-ai-roadmap-v10.pdf

Gold, H., Faqiri, H., Yeung, J., & Hu, C. (2023, October 8). Israel formally declares war against Hamas as it battles to push militants off its soil. CNN. Retrieved from https://www.cnn.com/2023/10/08/middleeast/israel-gaza-attack-hostages-response-intl-hnk/index.html

SumithaK. (2023). "Exploring the Role of Artificial Intelligence in Improving Passenger Satisfaction in the Airline Industry: An Analysis of Customer Feedback and AI-Driven Solutions.". doi:10.5281/zenodo.7827975

Kashyap, R. (2019). Artificial Intelligence Systems in Aviation. In T. Shmelova, Y. Sikirda, N. Rizun, & D. Kucherov (Eds.), *Cases on Modern Computer Systems in Aviation* (pp. 1–26). IGI Global., doi:10.4018/978-1-5225-7588-7.ch001

Kulida, E. L., & Lebedev, V. G. (2020). About the Use of Artificial Intelligence Methods in Aviation. 2020 13th International Conference "Management of large-scale system development" (MLSD), Moscow, Russia, 2020, pp. 1-5, doi: 10.1109/MLSD49919.2020.9247822.1-5

Kumar, M. (2022). Optimized application of artificial intelligence (AI) in aviation market. *International Journal of Recent Research Aspects*, *9*(4).

Lee, J. D., & See, K. A. (2018). Trust in Automation: Designing for Appropriate Reliance. *Human Factors*, *50*(1), 97–110.

Liang, Y., Wang, X., & Hou, J. (2021). Understanding User Needs for Artificial Intelligence in Aviation: A User-Centered Design Perspective. *IEEE Access : Practical Innovations, Open Solutions*, *9*, 1838–1847.

Liu, Y., Li, Y., Wang, G., & Dong, Z. (2020). A Comprehensive Review of Predictive Maintenance: Issues and Challenges. *Applied Sciences (Basel, Switzerland)*, *10*(8), 2847.

Merlo, T. R. (2024). *AI and data analytics applications in organizational management*. IGI-Global.

Nothing artificial about this intelligence: Delta's industry-first machine learning platform minimizes customer inconvenience during tough operations. (2020). Delta. Retrieved October 4, 2023 from https://news.delta.com/nothing-artificial-about-intelligence-deltas-industry-first-machine-learning-platform-minimizes

Patterson, R. E., Broy, N., & Leiden, K. L. (2022). User Empowerment in the Age of Artificial Intelligence: Designing for User Training and Interaction in Complex Systems. *Human Factors*, *54*(1), 5–24.

Ranganathan, G. (2023). Airline CEO's AI system for driving personalization. *Journal of Revenue and Pricing Management*, *22*, 166–170. https://doi.org/10.1057/s41272-022-00402-w

Saballa, J. (2023, October 13). Israel military using new AI system in combat operations: report. The Defense Post. Retrieved from https://www.thedefensepost.com/2023/07/17/israel-military-ai-system/

Samatas, G. G., Moumgiakmas, S. S., & Papakostas, G. A. (2021, May). Predictive maintenance-bridging artificial intelligence and iot. In 2021 IEEE World AI IoT Congress (AIIoT); pp. 413-419. IEEE.

Smith, A. M., Hancock, P. A., & Clegg, B. A. (2019). Task Load Management: Implications for Human-System Interaction. *Applied Ergonomics*, *74*, 234–242.

Stanton, I., Munir, K., Ikram, A., & El-Bakry, M. (2023). Predictive maintenance analytics and implementation for aircraft: Challenges and opportunities. *Systems Engineering*, *26*(2), 216–237.

Wilson, J., Sachdev, S., & Alter, A. (2016). How Companies Are Using Machine Learning to Get Faster and More Efficient. Harvard Business Review. Retrieved on October 4, 2023 from: https://hbr.org/2016/05/how-companies-are-using-machine-learning-to-get-faster-and-more-efficient

Wilson, A. D., & Russell, D. M. (2019). Human Performance with Control Room Automation: A Literature Review and Meta-Analysis. *Applied Ergonomics*, *76*, 192–204.

## ENDNOTE

[1] Hamas is a religious-based political movement. Hamas was formed in response to the escalating Zionist Israeli invasion of Palestine. The ideology espoused by Hamas is the ideology of Islam. Bachtiar (2009, p. 206-212).

# Chapter 3
# Aviation Cybersecurity:
## Shielding Against Cyberthreats in the Air

**Caner Asbaş**
https://orcid.org/0000-0002-2315-565X
*Atılım University, Turkey*

**Şule Erdem Tuzlukaya**
https://orcid.org/0000-0001-8244-6396
*Atılım University, Turkey*

## ABSTRACT

*Cybersecurity is a set of applications specialized for information and communication technology systems. Cybersecurity practices are conducted in cyberspace, and should be conducted both under the limitations and with the opportunities of this complex environment. Dynamic, totalistic, and adaptive approaches are required for cybersecurity; solely technical or financial measures would not be enough to ensure cybersecurity. Instead, a more holistic view that considers all aspects of cybersecurity would be more appropriate. In the aviation sector, where technical aspects are dominant and sensitivities are exceptionally high, it can be directly noticed that the associated complexities are heightened even further. This results in a requirement for a more concrete and detailed understanding and approach for cybersecurity in aviation. In this chapter, cybersecurity approaches and applications in aviation sector will be discussed extensively.*

DOI: 10.4018/979-8-3693-0732-8.ch003

## 1. INTRODUCTION

Cybersecurity can be defined as the specific set of applications focused on and specialized for information and communication technology systems, units, products and services. It is an interdisciplinary field based on many areas including computer science, law, psychology, management, transportation, and banking and finance. The cybersecurity paradigm, emerging from sociological, psychological and technical sources, can be embraced as a security-based perspective within the realm of information management.

Cybersecurity applications are conducted in cyberspace, which is defined as "a global domain within the information environment consisting of the interdependent network of information systems infrastructures including the Internet, telecommunications networks, computer systems, and embedded processors and controllers" (NIST, 2021) and it should be carried out both under the constraints and with the advantages of this complex environment. Therefore, with the increase in number of cyberattacks, cybersecurity applications require dynamic and adaptive approaches to overcome these complexities. On the other hand, this situation brings about the fact that cybersecurity cannot be guaranteed through solely technical or financial measures. Instead, a more totalistic approach that addresses all aspects of cybersecurity is considered appropriate.

In the aviation sector, where technical aspects dominate and concerns related to human life encompass both civil and military aspects, alongside the heightened sensitivities involved, it becomes evident that the associated complexities are even higher. This results a requirement for a more concrete and detailed understanding and approach for cybersecurity in aviation. In this chapter, cybersecurity approaches and applications in aviation area will be discussed extensively.

## 2. MAIN FOCUS OF THE CHAPTER

### 2.1 Cybersecurity and Cyberspace

The specific and specialized set of applications and implementations performed in order to ensure the security of systems, data, information, products and services and all other similar elements in connection with information and communication technologies or in the virtual environment, to protect them against external factors and to provide them with reliability and continuity are called cybersecurity. As an interdisciplinary field, cybersecurity can be regarded as both a field of practice and set of policies stemming from engineering, sociological, psychological and legal foundations (Luo & Zhdanov, 2016; Warkentin, Johnston, Shropshire & Barnett, 2016). The interdisciplinary scope of cybersecurity covers a wide range, including computer science, law, psychology, transportation, management, banking, and finance. This necessitates the consideration of many factors simultaneously and in a coordinated and harmonious manner.

The five pillar of cybersecurity paradigm can be listed as confidentiality, integrity, availability, non-repudiation, and authenticity or authentication (Porche, 2020). The explanations of these pillars are given in Table 1.

In this context, the cybersecurity lifecycle comprises three phases: preventing, eliminating, and reporting cyberattacks (Porche, 2020). These phases do not necessarily follow a linear path. For example, "the eliminating cyberattacks" phase may not always be achieved, and the "preventing cyberattacks" phase might remain incomplete, depending on the level of success in ensuring cybersecurity.

Since the main goal and reason of the existence of cybersecurity is the protection of cyberassets, all activities related to cybersecurity are carried out in cyberspace, considering the possibilities, opportunities, limitations, and constrains of cyberspace. From this perspective, cybersecurity can be examined in two different layers. In the narrower scope, only the five pillars of cybersecurity are taken as the focus for policies and implementations (Cavelty, 2015), while in the broader

*Table 1. Five pillars of cybersecurity*

| **Confidentiality** | **Keeping data and information secret** |
|---|---|
| Integrity | Preventing unauthorized data change and modifications |
| Availability | Readiness, accessibility and addressability of data when needed |
| Non-Repudiation | The declared and undeniable roles of receiver and transmitting parties in data traffic |
| Authenticity (Authentication) | User verification requirement and preventing unauthorized access |

context, all activities in cyberspace are evaluated under the heading of cybersecurity if they are related or linked to security goals or applications (Cornish, Hughes & Livingstone, 2009).

At this point, cyberspace is "a global domain within the information environment consisting of the interdependent network of information systems infrastructures including the Internet, telecommunications networks, computer systems, and embedded processors and controllers" according to the definition made by National Institute of Standards and Technology, U.S. Department of Commerce (NIST, 2021). The concept of cyberspace and related environments, which was included in the systematization due to the inability of the traditional classification of land, sea and air (in some cases, also space) used in warfare literature frequently, to position the virtual world (Even & Siman-Tov, 2012), has become a medium surrounding the physical and real layers in parallel with the rapid and effective advances in technology, and its deep integration to every aspect people's lives (Dyson, Gilder, Keyworth & Toffler, 1996).

Cyberspace, which is a composition of real, virtual, social and human factors and elements, can also appear primarily as a set of ideas, objects, subjects, structures systems and views (Deibert & Rohozinski, 2010). These elements can be grouped into three separate classes, the human layer, the logical layer, and the physical layer (Even & Siman-Tov, 2012). Among these classes, the human layer pertains to individuals using computer and other information and communication technology products and services. The logical layer represents to software, the Web, the Internet, and other similar elements devoid of physical existence, while the physical layer includes all infrastructure, systems, and hardware like cables, devices, and stations facilitating the connection and communication between the logical layer and the human layer.

Cyberattacks pose the most vital threat to the security of cyberassets within the realm of cyberspace. Cyberattacks, which target backdoors and security vulnerabilities that are common in almost every information and computer system, network and infrastructure, can be defined as an incursion or unwelcome interference carried out by cybercriminals, aimed at stealing, corrupting, deleting data or information, or rendering it unusable via information and communication technology products and services (Asbaş & Erdem Tuzlukaya, 2022). One of the factors that allows such an attempt to be called a cyberattack is the intent factor. Conversely, attempts without malicious intentions, unlike cyberattacks,

*Table 2. Fundamental types of cyberattack methods*

| | |
|---|---|
| Malware | Viruses |
| | Worms |
| | Trojans |
| | Ransomware |
| | Spyware |
| | Adware |
| Phishing | Whaling |
| | Pharming |
| Man-in-the-Middle (MitM) Attack | Session Hijacking |
| | IP Spoofing |
| Denial-of-Service (DoS) Attack | Distributed Denial-of-Service (DDoS) Attack |
| | TCP SYN Flood Attack |
| | Ping-of-Death Attack |
| | Botnet |

are called as cyberbreaches, even if they yield similar and comparable outcomes and progression (Ghadiminia, Mayouf, Cox & Krasniewicz, 2021; Mayo & Snider, 2016; Sommer & Brown, 2011).

Cyberattacks, which may be conducted to fulfill criminal, political, or personal aims, can employ a variety of methods (Peng et al., 2013). These methods can include applications that require a high degree of technical knowledge and experience, as well as approaches that are not directly based on computer science principles and techniques, such as social engineering, or the combination of technical and nontechnical approaches. Some fundamental types of cyberattack methods with their main subtypes are given in Table 2.

Cyberattacks have significant characteristic features that make them more appealing than other attacks. As a result, that the tools needed to organize a cyberattack are typically just one or more computers, access to networks such as LAN or the Internet, and specific software, often available for free or developed by the cyberattackers themselves. These features can be listed as cheapness, ease, availability, accessibility, anonymity and attacker-defender asymmetry (Karabacak, Yıldırım & Baykal, 2016). These features are presented in Table 3 with explanations.

*Table 3. Characteristic features of cyberattacks*

| Tools | Computer, network access, software |
|---|---|
| Cheapness | Cheaper than tools needed for other attack types |
| Ease | Easy to use |
| Availability | Simple to obtain |
| Accessibility | Generally, no permission required to obtain |
| Anonymity | User identity can be hidden easily |
| Attacker-defender asymmetry | Only single attack surface is required to organize a cyberattack, all attack surfaces should be protected to ensure cybersecurity. |

Due to these advantageous features, the defender's overall cybersecurity system becomes mutually dependent, interdependent, and complex (Chan, Yung, Lam, Tam & Cheung, 2001; Cichonski, Millar, Grance & Scarfone, 2012), presenting a challenging situation that should be handled holistically (Hathaway et al., 2012). It is an undeniable fact that the situations in a sector such as aviation, where technical aspects outweigh other factors, standards are complicated, and the human factor is at the forefront, are much more complex than depicted.

These mentioned advantages of cyberattacks have led to the idea that cyberattacks can be used as a complement to the warfare or as a separate battlefield, and the concept of cyberwarfare has entered the literature (Even & Siman-Tov, 2012). Thus, cyberspace has begun to be militarized, akin to land, sea, air, and even space. Additionally, given that air and aviation are already significant areas of militarization, the shift occurring in cyberspace, where aviation applications are closely interconnected and benefit from many opportunities, has doubled the importance of cybersecurity in the field of aviation.

As the biggest sources of threats to the cybersecurity just after cyberattacks, cyberwarfare can be defined as a set of cyberattacks to cause damage to the state which is regarded as enemy or organizations affiliated with this state at a scale comparable to the conventional warfare by using computer systems, computer infrastructures or computer networks against similar assets owned by rival state (Asbaş & Erdem Tuzlukaya, 2023). Cyberwarfare, driven by motivations such as military, ideological, terrorist, proxy war or hacktivism, encompasses various objectives, including espionage, counterespionage, propaganda, and sabotage-especially significant in the context of aviation. The critical infrastructure, facilities and services in the country targeted in sabotage, which is one of the most basic and relatively easy ways to case war-like damage in the hostile country. As potential targets, military systems, financial infrastructures, critical databases, flight control systems, energy grids, emergency services, healthcare systems, telecommunication networks, stock exchange systems, government services, factories, logistics systems, transportation systems, and airports represent these elements (Abele-Wigert & Dunn, 2006; Cavelty, 2015; Condron, 2007; Corbin & Strauss, 2008; Even & Siman Tov, 2012; Karabacak et al., 2016; Sridhar, Hahn & Govindarasu, 2012; Van Bossuyt & O'Halloran, 2019).

The aviation sector can be directly or indirectly negatively affected by this array of potential targets. At this point, many countries have identified aviation and especially civil aviation as a main asset for national security. For instance, the United Kingdom acknowledged the risks posed by cyberthreats to civil aviation in a 2012 report from the UK Center for the Protection of National Infrastructure. In 2021, Canada emphasized the importance of cybersecurity for computerized air traffic control systems in a report published by Canadian Security Intelligence Service. The United States established the Comprehensive National Cybersecurity Initiative (CNCI) in 2008 to ensure the cybersecurity of critical infrastructure, including aviation systems. Similarly, in 2001, the European Union recognized the challenges posed to aviation and airlines due by cyberthreats through the Commission of the European Communities (CEC) (Iasiello, 2014).

## 2.2 Cybersecurity methods in aviation

One of the most basic and simple ways to provide security against cyberthreats is the use of cyber-defense technology tools. These tools have been developed specifically against one or more specific types of cyberattacks and are efficient methods for providing a certain degree of cybersecurity. However, these special designs also bring some security deficiencies (Ghadiminia et al., 2021). Table 4 provides information about the basic cyber-defense tools and the types of cyberattack they are active and cannot provide security.

In addition to specialized technical tools, many guiding documents, development programs, and legal regulations about cybersecurity or related

*Table 4. Fundamental cyber-defense technology tools*

| Tool | Advantageous | Disadvantageous |
|---|---|---|
| Firewall | Unauthorized logins (network firewalls)<br>XSS attacks (application firewalls)<br>Cross-site forgery (application firewalls)<br>SQL injections (application firewalls) | Insider attacks (application firewalls)<br>Abusive passwords (network firewalls)<br>Unauthorized protocols (network firewalls) |
| Antivirus software | Malware<br>Viruses<br>Trojans<br>Worms<br>Keyloggers<br>Browser hijackers<br>Rootkits,<br>Spyware<br>Botnets<br>Adware<br>Ransomware | Against software not marked as malware in their database |
| Network proxy | DoS attacks<br>DDos attacks<br>MitM attacks | Accessed malicious software |
| Role-Based Access Control (RBAC) | Unauthorized access | Password misuse |
| Intrusion Detection and Protection Systems (IDS/ IPS) | Unauthorized access<br>Malware<br>Phishing<br>Rootkits<br>Trojans | Zero-day attacks |

issues have been published to prevent cyberthreats and ensure safety of critical organizations. Some examples of these documentation can be listed as the standards such as IEC 62443 and ISO/SAE 21434, program such as Capability Maturity Model Integration (CMMI), legislations such as General Data Protection Regulation (GDPR) by European Union, and frameworks such as Control Objectives for Information and Related Technology (COBIT) by Information Systems Audit and Control

Association (ISACA), ISO 27000 series by the International Organization for Standardization (ISO), System and Organization Controls (SOC) by the American Institute of Certified Public Accountants (AICPA) and Framework for Improving Critical Infrastructure Cybersecurity by the National Institute of Standards and Technology, US Department of Commerce (Bozkuş Kahyaoğlu & Çalıyurt, 2018).

Cyberthreats in the aviation sector can lead to security problems in different areas. These include air traffic flow, airport management and applications, as well as the operations of institutions and organizations involved in aviation related services. Having problems in aviation-related products or maintenance, as well as problems with reliability, represents a significant risk that can directly impact human lives and lead to significant financial losses (McAndrew, Vishnevskaya & Johnson, 2019). However, when cyberwarfare's purposes, especially the purpose of sabotage, is taken into account, facilities producing aviation-related equipment and vehicles, subcontractors and service providers working in connection with them, and their supply chains are attractive targets for the executives of cyberwarfare. As part of the growing importance in this context, Critical Infrastructure planning has become fundamental for all new equipment and software in the field of aviation (McAndrew et al., 2019). In addition to Critical Infrastructure Approach, one of the additional recommendations in the literature involves the utilization of artificial intelligence, especially artificial neural network modelling and implementations (McAndrew et al., 2019).

Airports are also vulnerable to cyberattacks and cyberthreats, both in terms of their management and flight control systems. This vulnerability may be in airports' internet infrastructure, telecommunication systems, and their own automation systems and software, as well as in newly used Internet-of-Things-based management and control infrastructures. As a result of cybersecurity defects, airports may experience problems in four basic areas, namely operational-maintenance, operational-efficiency, operational-management and services-augmentation (Koroniotis, Moustafa, Schiliro, Gauravam & Janicke, 2020). These problems can cause a wide range of problems, from a decrease in service quality to service interruption, disruption of aircraft traffic or even a complete stop. In the literature, the use of self-learning systems with artificial intelligence - machine learning methods has been recommended as an additional precaution to traditional cybersecurity methods (Koroniotis et al., 2020).

As mentioned above, it was the intent factor that differentiated a cyberattack from cyberbreach. In this context, being able to perceive user intentions, regardless of navigation, airport management or aviation-related production sector, will be an important advantage to detect and eliminate cyberthreats before they cause problems. In this direction, one of the methods suggested in the literature is CyberBERT architecture (McDonnell, Nada, Abid & Ehsan, 2021). With this method, which can be explained as a Deep Dynamic-state Session-based Recommender System for Cyber Threat Recognition, cybersecurity through malware classification is followed by a natural language processing model called Bidirectional Encoder Representations from Transformers (Devlin et al., 2018) in aviation short online sessions, aiming to increase the level of safety (McDonnell et al., 2021).

One of the possible sources of cyberattack in aviation is airport automatic fuel supply control system. This system ensures fuel safety, improves supply continuity and reduces personnel dependency (Dazhi, 2017). For this reason, the cybersecurity of these systems is of high importance in preventing possible cyberattack sabotage. Cybersecurity risk assessment method of ICS based on attack-defense tree model has been proposed in the literature for the protection of this system (Wang, Ding, Sui & Gu, 2021).

One of the interesting points is unmanned aerial system cybersecurity because it is much more sensitive and open to attack. These systems operate with the coordination and cooperation of many software and hardware that communicate with each other. For this reason, it is necessary to shield them against cyberthreats by providing cybersecurity in a way that the communication between them is confidential and encrypted. For this shielding, it is possible to find many encryption, codes and connection techniques in the literature, including blockchain applications (Kosuda et al., 2020).

Although cyber-defense technology tools can be effective against specific types of cyberattack, a more holistic and comprehensive approach is needed to ensure cybersecurity in a complex, complicated and error-free field such as aviation. This requirement can be met by applying strategic cybersecurity management principles to the aviation field. Strategic cybersecurity management can be defined as a framework that addresses the phenomenon and need of cybersecurity as a whole with a totalistic approach and allows the development of related strategies

(Asbaş & Tuzlukaya, 2022). This approach, which divides cybersecurity policies development into three stages as strategic cybersecurity planning, implementation and evaluation, has proposes some generic strategies, which can be adapted with respect to the application conditions, under the headings which are technical strategies such as implementing proactive or reactive defense, senior management strategies such as cybersecurity risk management and information sharing policies, human resources strategies such as conducting security education, training, and awareness (SETA) programs, legal strategies such as law-enforced surveillance, and organizational strategies such as reorganizing organizational structure, providing technological adaptability and flexibility, and developing organizational culture (Asbaş & Tuzlukaya, 2022).

When the subject is examined from this point of view, it will be revealed that cybersecurity practices in the aviation sector should also consider all aspects of the subject, including but not limited to management, organization, technical side, legal authorities and regulations and human resources, and developing an approach that will take into account all possible cyberrisks is required.

## 3. CONCLUSION

In addition to the benefits it provides, the integration of cyberspace into aviation also brings very dangerous cyberthreats to the agenda in this sector. Moreover, considering the importance of the aviation area in terms of logistics, supply chain, transportation and military, it is seen as an attractive critical area to be targeted for cyberwarfare as well. For this reason, the implementations of cybersecurity applications in the aviation sector and their success have become a vital cornerstone in the maintenance of the operations of this sector.

At this point, three different approaches can be mentioned for aviation cybersecurity. The first of these is the use of cyber-defense technology tools, as in other fields and in daily life. However, the limited protections provided by these tools do not provide the totalistic protection required by the aviation industry.

Second, there are some cybersecurity applications developed specifically for the aviation industry. Since their focus is aviation, they have a higher level of security compared to traditional cyber-defense technology tools, but they still offer customized solutions to single attacks or attack clusters instead of suggesting a holistic policy set.

Third, the application of the strategic cybersecurity management approach to the field of aviation comes. Since this approach considers cybersecurity from many points of view such as technical, social, human resources, legal, economic and organizational aspects at the same time, it can be examined more comprehensively and consistently in a very critical sector such as aviation, where reliability, continuity and maintenance are vital.

## REFERENCES

Abele-Wigert, I., & Dunn, M. (2006). The International CIIP Handbook 2006: An Inventory of Protection Policies in 20 Countries and 6 [*)*. Center for Security Studies.]. *International Organization, I.*

Asbaş, C., & Erdem Tuzlukaya, Ş. (2022). Cyberattack and cyberwarfare strategies for businesses. In F. Özsungur (Ed.), *Conflict Management in Digital Business: New Strategy and Approach* (pp. 303–329). Emerald Publishing. doi:10.1108/978-1-80262-773-220221027

Asbaş, C., & Erdem Tuzlukaya, Ş. (2023). Cyberwarfare: War activities in cyberspace. In F. Özsungur (Ed.), *Handbook of Research on War Policies, Strategies, and Cyber Wars* (pp. 128–145). IGI Global. doi:10.4018/978-1-6684-6741-1.ch007

Bozkuş¸ Kahyaoğlu, S., & Çalıyurt, K. (2018). Cyber security assurance process from the internal audit perspective. *Managerial Auditing Journal, 33*(4), 360–376. doi:10.1108/MAJ-02-2018-1804

Cavelty, M. D. (2015). Cyber-security and private actors. In R. Abrahamsen & A. Leander (Eds.), *Routledge Handbook of Private Security Studies* (pp. 89–99). Routledge. doi:10.4324/9781315850986-10

Chan, A. P. C., Yung, E. H. K., Lam, P. T. I., Tam, C. M., & Cheung, S. O. (2001). Application of Delphi method in selection of procurement systems for construction projects. *Construction Management and Economics, 19*(7), 699–718. doi:10.1080/01446190110066128

Cichonski, P., Millar, T., Grance, T., & Scarfone, K. (2012). *Computer Security Incident Handling Guide: Recommendations of the National Institute of Standards and Technology*. NIST. doi:10.6028/NIST.SP.800-61r2

Condron, S. (2007). Getting it right: Protecting American critical infrastructure in cyberspace. *Harvard Journal of Law & Technology, 20*(2), 403–422.

Corbin, J., & Strauss, A. (2008). *Basics of Qualitative Research: Techniques and Procedures for Developing Grounded Theory* (3rd ed.). Sage Publications. doi:10.4135/9781452230153

Cornish, P. N., Hughes, R., & Livingstone, D. (2009). *Cyberspace and the National Security of the United Kingdom: Threats and Responses.* Chatham House.

Dazhi, H. (2017). *Post-evaluation research on the oil supply project of Yantai Chaoshui Airport* [Dissertation, Qingdao University], Qingdao, China.

Deibert, R. J., & Rohozinski, R. (2010). Risking security: Policies and paradoxes of cyberspace security. *International Political Sociology, 4*(1), 15–32. doi:10.1111/j.1749-5687.2009.00088.x

Devlin, J., Chang, M. W., Lee, K., & Toutanova, K. (2018). *BERT: Pre-training of deep bidirectional transformers for language understanding.* In *Proceedings of NAACL-HLT 2019*, Minneapolis, MN, US.

Dyson, E., Gilder, G., Keyworth, G., & Toffler, A. (1996). Cyberspace and the American dream: A Magna Carta for the knowledge age. *The Information Society, 12*(3), 295–308. doi:10.1080/019722496129486

Even, S., & Siman-Tov, D. (2012). *Cyber Warfare: Concepts and Strategic Trends.* Institute for National Security Studies.

Ghadiminia, N., Mayouf, M., Cox, S., & Krasniewicz, J. (2021). BIM-enabled facilities management (FM): A scrutiny of risks resulting from cyber attacks. *Journal of Facilities Management.* doi:10.1108/JFM-01-2021-0001

Hathaway, O. A., Crootof, R., Levitz, P., Nix, H., Nowlan, A., Perdue, W., & Spiegel, J. (2012). The Law of Cyber-Attack. *California Law Review, 100*(4), 817–885.

Iasiello, E. (2014). A review of "a framework for aviation cybersecurity" – a decision paper from the American Institute of Aeronautics and Astronautics. *Journal of Homeland Security and Emergency Management, 11*(1), 95–100. doi:10.1515/jhsem-2013-0076

Karabacak, B., Yıldırım, S. Ö., & Baykal, N. (2016). A vulnerability-driven cyber security maturity model for measuring national critical infrastructure protection preparedness. *International Journal of Critical Infrastructure Protection*, *15*, 47–59. doi:10.1016/j.ijcip.2016.10.001

Koroniotis, N., Moustafa, N., Schiliro, F., Gauravaram, P., & Janicke, H. (2020). A holistic review of cybersecurity and reliability perspectives in smart airports. *IEEE Access : Practical Innovations, Open Solutions*, *8*, 209802–209834. doi:10.1109/ACCESS.2020.3036728

Kosuda, M., Lipovsky, P., Yang, W., Bai, M., Novotnak, J., & Szoke, Z. (2020). *Discussion on blockchain applications in unmanned aerial systems domain. 2020 New Trends in Aviation Development*. NTAD., doi:10.1109/NTAD51447.2020.9379109

Luo, X. R., & Zhdanov, D. (2016). Special issue introduction: A comprehensive perspective on informationsystems security — technical advances and behavioral issues. *Decision Support Systems*, *92*, 1–2. doi:10.1016/j.dss.2016.10.003

Mayo, G., & Snider, D. (2016). *Bas and cyber security: A multiple discipline perspective*. In *2016 International Annual Conference of the American Society for Engineering Management*, Concord, NC, US.

McAndrew, I., Vishnevskaya, E., & Johnson, M. (2019). *Artificial intelligence in the aviation manufacturing process for complex assemblies and components*. In 2019 6th International Conference on Mechanical, Materials and Manufacturing, Boston, MA.

McDonnell, S., Nada, O., Abid, M. R., & Amjadian, E. (2021). CyberBERT: A deep dynamic-state session-based Recommender System for Cyber Threat Recognition. *2021 IEEE Aerospace Conference*. IEEE. 10.1109/AERO50100.2021.9438286

NIST. (2021). *Cyberspace*. NIST. https://csrc.nist.gov/glossary/term/cyberspace

Peng, Y., Lu, T., Liu, J., Gao, Y., Guo, X., & Xie, F. (2013). Cyber-physical system risk assessment. In *2013 Ninth International Conference on Intelligent Information Hiding and Multimedia Signal Processing*. IEEE. 10.1109/IIH-MSP.2013.116

Porche, R. P. III. (2020). *Cyberwarfare: An Introduction to Information-Age Conflict*. Artech House.

Sommer, P., & Brown, I. (2011). Reducing Systemic Cyber Security Risk. *Organisation for Economic Cooperation and Development Working Paper No. IFP/WKP/FGS, 2011*(3).

Sridhar, S., Hahn, A., & Govindarasu, M. (2012). Cyber-physical system security for the electric power grid. *Proceedings of the IEEE, 100*(1), 210–224. doi:10.1109/JPROC.2011.2165269

Van Bossuyt, D. L., & O'Halloran, B. M. (2019). A method to choose between automation and human operators for recovery actions during a cyber attack. *Procedia Computer Science, 153*, 352–360. doi:10.1016/j.procs.2019.05.089

Wang, S., Ding, L., Sui, H., & Gu, Z. (2021). Cybersecurity risk assessment method of ICS based on attack-defense tree model. *Journal of Intelligent & Fuzzy Systems, 40*(6), 10475–10488. doi:10.3233/JIFS-201126

Warkentin, M., Johnston, A. C., Shropshire, J., & Barnett, W. D. (2016). Continuance of protective security behavior: A longitudinal study. *Decision Support Systems, 92*, 25–35. doi:10.1016/j.dss.2016.09.013

# Chapter 4
# Online Engagement: From Public Relations Perspectives

**Sevilay Ulaş**
*Çankaya University, Turkey*

## ABSTRACT

*The aviation industry can be said to be one of the industries whose star has been shining more and more recently. In its simplest form, can be defined as a process that includes all activities, details and industries related to aviation. Aviation industry institutions aim to carry out effective and sustainable communication activities with their target audience/customers and followers. Social media tools and environments are also the preferred media in this communication process. Social media influencers, which can be specified as an increasingly remarkable communication element, are also in the aviation field and their effective use is seen. Its applications are seen in many different areas such as aircraft-aviation, travel, food and beverage, destination. In the effectiveness of the communication process, online engagement is also a factor that is mentioned. Therefore, the two-way communication process that takes place in this study, which focuses on influencers in the aviation field and online engagement, is also mentioned in the specified communication process.*

## 1. INTRODUCTION

The place of change and transformation in information communication technologies in communication tools and usage practices draws attention. It is possible to talk about the existence of these digital-evolving communication processes in professional life practices as well as in daily life. Therefore, the existence of digital transformation is seen in

DOI: 10.4018/979-8-3693-0732-8.ch004

the ways of doing business, management practices and strategies in their institutions. Within the scope of this digital transformation, the concept of digital innovation takes its place. Digital innovation can also be described as a concept that has a key carrier role in this process that has turned into a kind of digital. With digital innovation, institutions can implement innovations that meet digital in many areas such as management practices and strategies, customer and employee relations, corporate communication processes and strategies. Reflections of digital transformations can be seen in almost every field of activity. One of these areas is the aviation industry. In this field of activity, the reflections of digital transformation and innovation practices are seen in both managerial and communication-related practices. Aviation management, by its nature, includes many elements and includes the processes of managing them. At this point, both the way the institutions do business, their internal communication activities, their innovative approaches and practices, and the communication activities they carry out with their stakeholders and target groups outside the institution are also important in their strategies. Especially in the institutional dimension, how the concept of innovation is implemented and how it is communicated is also an important dimension as well as its existence. The communication activities carried out inside and outside the corporations, the communication tools and elements used, and how they can keep up with innovation are among the points that should be focused on. The importance and effective role of communication in almost every field of activity cannot be denied, as well as in the aviation management process. The impact of digital innovation in the focus of institutions or aviation management is remarkable. The reflection and role of these digital innovations in the communication processes with the target audience and stakeholders is also an important detail. These digital innovations take their place as an important component at the point of sharing and communicating with the target audience and establishing a connection with the target audience, which is important for institutions.

As mentioned before, the digital transformation process emerges as an inclusive concept that includes many different elements. In this process, applications within the scope of the quality of communication activities and innovative approaches in communication tools draw attention. Sharing the innovative applications with the target group(s) and stakeholders in these digital-based management and implementation processes makes their presence known in terms of bonding and corporate sustainability.

At this point, influencers, which have been showing themselves in many fields of activity recently and with which institutions cooperate as an effective communication tool, draw attention as a shining factor. From this point of view corporations exist as a communication element especially in these communication processes. In this process, which has evolved into digital with almost every component, there are of course changes and transformations in the quality of the communication process. In this sense, in addition to social media influencers who take part in a digital communication process, there is also online engagement as another element. It is possible to talk about the existence of the engagement factor, which is created and tried to be maintained on online platforms as the communication quality and feature mentioned. Online engagement, which is aimed and created with the communication activities carried out in this communication process, which proceeds on a digital basis, is an effective factor. Therefore, in addition to cooperating with social media influencers in the communication processes of institutions with their target audience and followers, online engagement is also important. The online engagement factor and rate, which is one step beyond mere communication and is an element on the basis and continuity of communication, also plays an important role in the quality of the communication process. Since the recent past, the presence and effectiveness of digital communication elements come into prominence with the transfer of communication activities of institutions with their target audience/followers and stakeholders to a digital base. Social media influencers are in addition to other elements involved in online engagement, bonding with target audience/followers and maintaining them in the corporate communication process. Influencers create an impact with the content they share on different social media applications. They influence and guide their followers in areas such as shopping, fashion, health, cosmetics and culture. They appear to be digital opinion leaders while guiding and influencing their followers in their field of expertise (Allianz.com.tr). Influencers, who are a kind of opinion leaders among their followers, have an impact on their followers with their posts. From this point of view, it is seen that influencers are also effective in daily life practices and styles.

In this study, it is aimed to reveal the relationship between social media influencers and online engagement. At this point, it is tried to describe how an online engagement is created with the content shared in the focus of the examples of influencers in the aviation industry and its reflections in the aviation industry. In this context, the two-way symmetrical communication perspective of public relations will be utilized in the online engagement process. In this way, it is planned to describe how the online engagement element, which can be accepted two-way in nature, takes place in this sector through social media aviation influencers. For this purpose, content sharing of social media influencers active in the aviation industry through their official Instagram accounts will be evaluated. Through these examples, it will be tried to define how influencers take place as an innovative element in the aviation field. Accordingly, of course, it is planned to make an evaluation within the concept of online engagement, which emphasizes the effectiveness of communication. In this way, as a new communication element, it is aimed to define the role of social media influencers in the aviation sector with their presence and communication activities. The influencer samples in question will be determined among the social media influencers who are present and actively involved in the aviation field.

Social media influencers are among the fields of study that attract attention with their application examples in professional life as well as in the academic field. They are among the current topics of research, especially in the academic field. Examples of social media influencers in the field of activity that this study focuses on, and application practices are included. As stated above, it is increasingly seen as a field of study that attracts attention and has efficiency. From this point of view, the work planned in the focus of social media influencers in the aviation field within the main purpose of the study will contribute to the field in question and lead to future related studies. Influencers are seen as effective communication tools to influence and connect with their followers and target audiences.

## 1.1 Aviation Industry and Aviation Influencers

Aviation industry emerges as a sector that includes many different fields of activity. It can be said that the aviation industry is one of the sectors that has been growing and developing relatively rapidly in recent years. Especially within the scope of the growth in tourism and cargo demand worldwide, reflections and effects are also seen in the sector (Dube et al., 2021). Aviation applications can be seen in different ways in different geographies. As it mentioned (Hickie, 2006); there are many different regions of the world, for example Seattle and Toulouse as noted in the study (Broekel & Boschma, 2012). It is possible to talk about the existence of intensive investments in this field, especially aviation companies (Gisario et al., 2019).

Along with the growth and development in its field, has a positive effect on the passengers as it offers different destinations. At this point, the positive effects of each stakeholder in air transportation (airlines, airports, air traffic management (ATM), aircraft manufacturers, etc.) draw attention (Dožić, S. 2019: 1). The aviation industry can be described as a dynamic sector that constantly develops within the scope of its characteristics and dynamics. It may encounter a number of different requirements and changes in the long and short term (Linz, M. 2012: 28). These changes and developments draw attention in almost every field. In addition to the services offered by the aircraft industry, production details are also important. At this point, it is involved in studies related to recycling (Wong et al., 2017). Therefore, the aviation industry has a structure that develops itself and focuses on technologies. It takes its place in technologies that have the potential to change the way aircraft are designed, manufactured and repaired in the future (Wagner & Walton, 2016).

The materials used in the development and production processes of this industry are also important. At this point, a focus can be created on new materials that provide sufficient structural stability and are lightweight, depending on the need and situation (Ellrich et al., 2020). The production conditions of the aviation industry, the fossil fuels used, the emission characteristics and the details of environmental sensitivity appear as situations that need to be studied (Karaman et al., 2018). It draws attention in the technologies used and focused within the scope of these developments. High technology is mentioned in every field. (Perrow, 1984); the aviation industry incorporates high technology in working and production conditions, along with the view that 'accidents are inevitable' (Gill & Shergill, 2004).

In this sense, the air transport industry is one of the industries that has attracted increasing attention in recent years within the scope of the conditions and dynamics it is in. In this change and development process, the existing conditions such as the competitive environment, the demands and expectations of the customers, the tariff and the content of the services that exist with the industry in question also manifest themselves (Malighetti et al., 2011). However, in these developments and the difficulties that may be encountered, in addition to the production part, the management and managers also have a position and place. The determination of these situations can guide and benefit both the operation and the managers (Alam, M. A. 2016: 68). The aviation industry has evaluated many methods, techniques and procedures to reduce operational costs in certain periods and conditions (Weckman et al., 2001).

As seen in all these developments, the use of technology is crucial. In addition, digital transformation is seen in communication processes as well as in production. It draws attention in communication processes as well as in production and management processes. In the aviation industry, institutions in every field of activity strive to communicate effectively and continuously with their target audiences/customers. It is possible to say that the communication tools and elements used are also important at this point. Considering that digital transformation manifests itself in almost every field and sector, there are some evolutions in the aviation industry. Today, based on the understanding that global digital transformation is a reality that determines the development of businesses and markets, digital reflections are also seen in the aviation industry (Tikhonov et al., 2019). In addition to digital transformation, the aviation industry is faced with high production operating costs and intensive technology (Xiong & Wang, 2022).

With the meeting of internet-based applications and technology within the scope of the digital transformation process, it has led to changes both in the corporate dimension of the airline industry and in the communication of the same institutions with their customers and in the way of travel. In this context, airline companies can also use digital, internet-based applications in order to provide a better service (Avram, 2017: 63). It is important that internet applications and computers can be connected to each other. Passengers can make ticket reservations and online check-in through these applications and computers (Scott & Trimarchi, 2017).

It is possible to say that digital technologies are used with many different applications in the aviaiton industry. In the table below, there are explanations about which digital technologies are used and how at Table 1.

The digital transformation in the aviation industry causes airports to constantly renew their infrastructure. Infrastructure services needed to

*Table 1. Digital Technologies in aviation industry (Катерна, О. К., & Молчанова, К. М. 2020:58-59).*

| Digital Technology Name | Descriptions |
|---|---|
| Blockchain Technology | Airlines can use blockchain technology to improve their operational efficiency and customer experience. |
| Augmented Reality and Virtual Reality (AR/VR). | Airports can use AR to help passengers navigate the complex layout of the airport. |
| Artificial Intelligence (AI) | The application and use of AI makes it possible to provide each passenger with accurate information at the right time, based on location, time before flight, profile and preference. |
| Beacons Technology | Beacons technology allows passengers to navigate between different terminals at the airport. |
| Robotics | Airline industry uses robots to assist with various tasks, customer management, baggage handling, parking, etc. |
| Biometrics | The airline industry is constantly working to make the travel experience enjoyable and comfortable for its customers. |
| Big Data and Analytics | Airlines are helping to create pleasant experiences for passengers, understand customer preferences in real time and provide them with customized offers, etc. They can obtain insights by analyzing the large amount of data they have in order to present. |

support these modern services are listed below: (Катерна & Молчанова, 2020).

- Wireless Local Area Networks
- Automatic document readers connected to border control systems
- RFID detection infrastructures
- Biometric identification systems
- Electronic ticket readers
- Self-service kiosks
- Security system (Катерна & Молчанова, 2020).

The use of digital applications in aviation is a remarkable situation in terms of providing the connection between the air and the ground. An example would be a digital twin, expressed as a virtual copy of a physical asset such as an aircraft engine. So this element provides the opportunity to show the engineers on the ground how the engine works while the aircraft is in the air (Ibrion et al., 2019).

It is possible to talk about a number of market figures for this sector, where digital transformation is experienced and application practices take place. The Aviation Market size is expected to grow from USD 333.96 billion in 2023 to USD 386.21 billion by 2028, at a CAGR of 2.95% during the forecast period (2023-2028). (https://www.mordorintelligence.com/industry-reports/aviation-market). The global aviation analytics market size is projected to grow, 2021 to 2028 from USD 2.02 billion in 2021 to USD 4.36 billion in 2028 at a CAGR of 11.58% in the 2021-2028 period (Fortunebusinessinsights.com, 2020).

Communication tools used in the digital transformation process are also noteworthy in terms of their effectiveness and application practices. Especially in recent years, effective, robust and interactive communication processes with the target audience/customers have also gained importance. Social media tools and environments can be specified as one of these communication media. Application practices in social media tools and environments are also seen in the tourism industry. Communication activities within the scope of the tourism industry are also reflected in customer behavior and practices (Bigne et al., 2018). Due to its nature, social media is used both in the field of tourism and in communication processes with the target audience, as it allows for two-way and effective communication. There are examples of effective communication with consumers in many different institutions and fields of activity within the tourism industry (Hays et al., 2013).

On the other hand, social media tools and environments have enabled travelers in this sector to share their travel experiences quickly and easily in communicating with the target audience. Therefore, information shared on social media sites is considered an important source of information that can affect travel decisions for potential travelers (Kang & Schuett, 2013). At this point, it is seen that social networks are effective in establishing and maintaining communication with customers/followers. In social media platforms, an engagement can be created with content, information sharing and the participation of customers/followers (Punel & Ermagun, 2018). Organisations can use different communication tools at the point of creating interaction with their target audiences in their communication activities. They pay attention to factors such as the details of the content created and shared, being new and different, in reaching their target audiences/followers, ensuring continuity and creating engagement (Knoblich et al., 2017). Through social networks, institutions can bring their products and services together with their target audience/followers, communicate, and make arrangements with the feedback received from them. These elements are also important in the perception and image of the institution / brand that is formed and created for their institutions (E. Smith et al., 2023). Information, opinions and recommendations obtained from social media environments are becoming more important for potential customers/travelers. Therefore, social media tools and environments are positioned as an effective communication medium and tool in terms of sharing their experiences and information, and it can be said that the communication provided in this way also plays a role in the preference for the institution/brand (Pop et al., 2022).

Instagram, one of the social media applications, has become popular in recent times, especially with content sharing in this field. This platform has become the go-to tool for those who are looking for unique travel inspiration, where one can share their travel and aviation adventures, personal experiences (Influencermarketing.ai. 2023). Airline companies can share their content from their official accounts and create engagement. Content shared from their official accounts has a role in the degree of impact (Wibisono, A. I., & Ruldeviyani, Y. 2021: 318). While the content shared in social media applications contributes to the activities of the institution such as promotion and advertisement, on the other hand, it also plays a role in establishing bonds with target audiences and connecting customer groups to each other (Binthiyaa & Fazıl, 2023).

It is safe to say that digital transformation / adaptation is experienced in almost every sector and field of activity. Aviaiton industry is one of them and it realizes social media and digital applications within the scope of internet-based applications in the digital transformation process.

## 1.2 Social Media Influencers and Online Engagement

Social media influencers draw attention as a preferred communication tool/element in communication planning, which is involved in social media tools and environments and in the application practices of institutions. The categorization and explanations for social media influencers can be seen in the table below, Table 2.

Details on the categories, advantages and disadvantages for influencers that have different categories within themselves are given in the table below, Table 3.

*Table 2.Types of social media infleuncers (Campbell, C., & Farrell, J. R. 2020: 471-472).*

| Categories | Definition |
|---|---|
| Celebrity influencers | A celebrity influencer is any individual who enjoys public recognition outside of social media and is leveraged by brands for their large follower base. |
| Megainfluencers | It can be expressed as a everyday celebrity.<br>It can be characterized as people who experience a significant increase in followers on social media and reach 1 million or more followers. |
| Macroinfluencers | It can be expressed as a sweet spot.<br>They are collaborative influencers who have between 100,000 and 1 million followers, can reach strong participation rates, can be used for brand display/ promotion. |
| Micro | It can also be expressed as a rising star.<br>They are also smaller in scope than macro-influencers. Audiences tend to be more localized according to their geographical basis. |
| Nano | They can be expressed as newcomers.<br>Influencers are at the beginning of their careers and followers are often described as friends, acquaintances, and others living nearby. |

*Table 3. Definition and characteristics of micro, macro and mega-influencers (Authors' creation, based on previous literature and industry reports (Campbell & Farrell, 2020; Moffitt & Azarfar, 2021; Porteous, 2018; Sinkwitz, 2020 – Akt: Conde, R., & Casais, B. 2023:2).*

| | Micro-influencers | Macroinfluencers | Mega-influencers |
|---|---|---|---|
| Followers > | 1,000 – 100,000 | 100,000 – 1,000,000 | 1,000,000 |
| Characteristics | Has a strong voice in a community or niche. They are perceived as highly trustworthy by their followers. | There is a significant number of loyal followers. | They can become famous on the Internet and become more well-known on a global scale. |
| Advantages | Elements such as loyal audience, originality. | Easier to communicate at the point of accessibility. | High visibility as part of reaching the masses. |
| Disadvantages | There may be lower reach and visibility. | Engagement rates can be said to be around 5 to 25%. | There may be relatively less engagement. |

Engagement/online engagement is a striking element in the practice practices of social media influencers or aviation industry influencers in the aviation industry. Online engagement is an important and important element in influencer practices, especially in social media applications. In this view, as participation and engagement, which is one of the most important concepts that are considered and attract attention in the communication activities carried out in social media tools and environments. From this point of view, it can be said that ''engage'' and/or ''engagement'' appear to replace more traditional relational concepts, including (Brodie et al., 2011).

The concept of engagement appears in traditional communication tools and environments as well as in new media tools and environments. Therefore, many different definitions or qualifications can be found. In studies published in the relevant literature and resources, interaction in practice is defined as a combination of cognitive aspects (for example, being interested in a company's activities), behavioral aspects (participating in the company's activities), and/or emotional aspects (feeling positive about a company's activities) (Dijkmans et al., 2015). It is possible to say that this concept, which has made a name for itself in the traditional environment, has also been adapted to digital and used. The concept of Engagement also exhibits a situation with different definitions and components in itself. The mentioned interaction takes its place as a situation created with customers/followers. At this point, customer loyalty can also be described as a psychological state that emerges through engagement with an institution/brand and creative customer experiences with the institution/brand and customers/followers. The levels and impact rates of this engagement may differ (Brodie et al., 2011).

The use of social media platforms, the implementation practices realized, and the online communication between the institution-target audience/followers are effective for sharing their experiences (Tian et al., 2020). Online engagement has become widespread with the communication activities carried out in social media tools and environments. In this point, the situation of online participation in producing and sharing content may differ according to time and place (Wang & Liu, 2019). When consumer engagement is considered as a context, it appears as a process that includes many different dynamics. In purchase intention and behavior and beyond, interaction and communication activities seem to be a phenomenon that relates to and can reveal brand communities (Dessart et al., 2015). Within this boundries, customer experiences have been transformed along with the usage practices of digital communication technologies in almost every field. It is also seen that the online engagement created in this way has turned into an important aspect of the modern customer experience in the hospitality industry (Kim et al., 2020).

*Table 4. Example of social media engagement manifestations (Dessart, L., 2017)*

| Dimensions of engagement | Brand focus | Community focus |
|---|---|---|
| Affective | A happy customer / consumer who gets an answer to a question customer asks about a brand in the social media environment. | A consumer who enjoys interacting with other fans of a brand on social media. |
| Cognitive | A consumer caught up in content posted by the brand on social media. | A consumer who pays great attention to the comments and responses of other consumers about the brand on social media. |
| Behavioural | A consumer who shares her opinion about a product with the brand on social media. | A consumer seeking information about the brand and seeking help or advice from other members of the community on social media. |

Social media activities such as viewing, commenting and even sharing content may not be the same as engagement and may not have the same meaning. The concept of engagement can be characterized as a state of mind and emotion and a different level of participation in addition to and beyond such communication activities (Smith & Gallicano, 2015). Social media users have different engagement on the relevant platforms (Paek et al., 2013). There are some different qualifications for the concept of engagement in social media tools and environments. Table 4 provides examples of consumer engagement on social media for each dimension and focus (Dessart, L. 2017).

Current events also play a role in the motivation and awareness of corporate/corporate brands to follow the content of social media followers, which have interaction in their natüre (Alraies & Sahni, 2017). Many airline brands are looking for different and effective ways of communication in order to be able to communicate effectively with their target audience and to gain more passengers, and related strategies that focus on the emotions of their company/brand and establish a handle. At this point, the communication activities carried out in social media tools and environments (such as producing content, responses from followers to content, feedback, listening to customers/followers) play an important role (Chumwatana & Chuaychoo, 2017).

In additon, social media influencers in the Aviaiton industry are the communication tools/elements that are actively preferred and increasingly shining in the mentioned communication purposes and applications. It is seen that these influencers, which are also described as social media influencers or, depending on the situation, digital influencers, affect the followers and have a role, especially at the point of e-wom, together with elements such as their shares, content, advice / promotion. At this point, it can be said that platforms such as Instagram and YouTube have taken their place as popular platforms due to their wide reach and focus on visual content. Therefore, it can be said that they use the said platforms in the tourism and airline industry, in cases related to their field, to share with their customers / followers, together with influencers (Barbe & Neuburger, 2021).

As a field, social media influencers are an application practice and an effective strategy in line with various objectives such as increasing customer contact, establishing bonds with them, creating engagement with them in various communication activities such as promotion, advertising and marketing together. In recent times, it is seen that customers/followers take into account the qualifiers of influencers that can be references or accepted by the institution. In this way, it can be said that it is evaluated in a preference for the institution (Wardani et al., 2019).

According to CreatorDen's "Overview 2023" report, which examines current data and developments on influencer marketing, 23 percent of the brands that have collaborated with influencers have collaborated with content producers for the first time in the last 12 months, and 54 percent have collaborated with content producers for the first time in the last 3 years. Among the social media platforms preferred in Influnecer collaborations, Instagram ranks first with a rate of 97.7% (marketingturkiye.com.tr, 2023). According to the research results, it is noteworthy that brands do not always prefer to work with the same content producers. Accordingly, 47 percent of the participants state that they "mostly work with different content producers" (digitalage.com.tr, 2023).

Data published by Influencer Marketing Hub shows that the influencer marketing market, which was 1.7 billion dollars in 2016, increased to 9.7 billion dollars in 2020. According to another data shared by Influencer Marketing Hub, 90 percent of brands believe in the power of influencer marketing. 80 percent of marketers say the results are effective. 89 percent say it works just as well, if not better, than other channels; 71 percent say that the traffic and consumer quality from influencer marketing is better than other sources (marketingturkiye.com.tr, 2021).

Aviaiton industry social media influencers have recently produced relevant content within their own structure, as well as communicating with institutions and carrying out communication applications in the relevant sector.

## Social Media Influencers on Aviation Industry From Public Relations Perspectives

Public relations, by its nature, includes two-way communication. It will be evaluated within the scope of the two-way symmetrical communication process. When we look at the communication process established through social media influencers, public relations and symmetrical communication come together at one point, together with the two-way communication process made possible in the social media environment. The communication process carried out is considered as a two-way communication process with the basic point of view of public relations. The online engagement element that emerges in this way is also considered within this scope.

Shared content such as travel experiences shared on social media is also important for institutions and customers/followers in the field of tourism. With these shared content on social media, it is possible to interact with the institution and customers/followers (Kang & Schuett, 2013). With user-generated content (UGC), which is one of the most important features in the nature of social media, both the institution, the source, and the customer / followers can be active, produce content and create engagement. The existence of different and effective communication strategies can be made possible with these created contents, travel experiences, information about the institution, and interaction. It is possible to talk about the sharing of independent and reliable information and experiences through social media travelers and influencers who are active in social media. Therefore, it can be said that these shared contents play a resource role in creating communication/interaction strategies of institutions in areas such as tourism and airline industry (Kang & Schuett, 2013).

It is seen that Aviaiton influencers have categories related to different fields and fields within themselves. They may refer to different areas such as aircraft operators, travel experiences, content related to tourism. Therefore, reports containing rankings and evaluations for different influencer content from different fields are published. Within the framework of some of the reports in question, aviaiton is not for influencers.

Looking at the 2023, 06 2023 rankings of Top 20 Aviation Influencers, it is seen that different and similar content is produced and shared. The sequence proceeds as follows. Aviation influencers in the said list are Guillaume Laffon, Karim Nafatni, Instagram Handle @karimthecaptain takes a place. Then, Steveo comes in the list. Also, Instagram Handle @steveokineevo. Deon Mitton, Instagram Handle @deonmitton. While providing information about the flight, they also share their travel and flight experiences (Feedspot.com, 2023). When looking at other influencers in the aviation field, it ranks first

Biggest influencers in Airlines are followed (Airport-Technology.com, 2021). Isaac Alexander (@jetcitystar). Isaac Alexander is the chief editor of the Jet City Star blog, where he covers news and information on the aerospace and aviation industry. He previously served as aircraft engine mechanic for the US Navy where he worked as the powerplant mechanic on the GE F404 and F414 engines of the McDonnell Douglas F/A-18 Hornet and Boeing F/A-18E/F Hornet. Alexander previously co-founded and served as the director of publicity for the Maelstrom International Fantastic Film Festival. In the continuation another aviation influencer is named Alex Macheras (@AlexInAir). Alex Macheras is an aviation analyst who discusses global aviation industry news on international news networks such as CNN, BBC News, Sky News, Al Jazeera and ITV, among others. He collaborates with airline industry experts, leaders and policy makers who have a global perspective on aviation and provide the latest aviation news from around the world.

Jon Ostrower (@jonostrower) is another aviation influencer. Jon Ostrower is editor-in-chief of The Air Current, a website that provides news and business insights about the aviation industry.

Fourth-placed Jason Rabinowitz (@AirlineFlyer) is another aviation influencer. Jason Rabinowitz is head of Airline Research at The Airline Tariff Publishing Company (ATPCO), which collects and distributes fare data for the airline industry. It specializes in airline amenities product, collecting data on airline passenger experience from hundreds of airlines. Tim Robinson (@RAeSTimR), aviation influencer, is editor-in-chief of AEROSPACE, the main journal of the Royal Aeronautical Society, the foremost organization striving for the advancement of aeronautical science and development. Robinson specializes in environmental science, transport and travel and has been featured in the BBC, CNN and Irish Times (Airport-Technology.com, 2021). It can be said that Instagram, an application dominated by social media, which manifests itself in almost every area of life, has become a platform for people/influencers to showcase their travel and aviation adventures. And when we look at the travel, tourism and aviation Instagram phenomena in Celle in 2023, below are the aviaiton influencers, respectively (Influencermarketing.ai., 2023).

Instagram, an application dominated by social media, which manifests itself in almost every area of life, has become a platform for people/influencers to showcase their travel and aviation adventures. And when we look at the travel, tourism and aviation Instagram phenomena in Celle in 2023, below are the aviaiton influencers, respectively (Influencermarketing.ai., 2023).

♛ Dennis ♛ – @dennis.herlitz, an Instagram phenomenon from Celle, Dennis has become one of the leading figures in the travel, tourism and aviation industry in recent years. From stunning landscapes to incredible flights, he has the ability to find the most awe-inspiring moments and share them with his followers. When we look at another influencer, Ricetime – @ricetime.celle is a modern travel inspiration. Ricetime is one of the most talented travel, tourism and aviation influencers around. He is an Instagram influencer who knows everything about travel, tourism and aviation. With the relatively high rate of interaction in her posts, she quickly creates a follower base of thousands of fans who can't get enough of her beautiful content. Third place on the list is Karina Pfeiffer – @ katinka_pe, Karina Pfeiffer is an Instagram influencer focusing on travel, tourism and aviation. Katrina Pfeiffer is an Instagram influencer with a strong interest in travel, tourism and aviation. Karina's Instagram account, which mainly features visually stunning travel photos, has a relatively impressive interaction rate with each post. Karina Pfeiffer, better known on Instagram as katinka_pe, is a popular influencer who enjoys sharing her adventures in travel, tourism and aviation (Influencermarketing.ai., 2023).

Within the scope of Internet-based applications, social media are widely used tools that have achieved significant global influence. It can be described as a platform that attracts more attention and arouses interest, especially with social media influencers (Wibisono & Ruldeviyani, 2021).

As stated above, aviation influencers produce content for the relevant field in line with their fields of activity. These consist of content such as travel, food and beverage, nature/nature activities, aircraft industry/

aviation. It can be said that interaction can be created by establishing an effective communication with the targeted audience/customer or followers with these applications that are carried out in general.

## CONCLUSION

The concept and process of public relations can be defined as a two-way communication process based on trust and mutual understanding, which a person or institution realizes with its target audience/public. This communication process can be carried out in different communication environments and with different tools. In recent years, there have been differences in tools and environments, especially with the changes in the field of communication. With the reflection of digital transformation on corporate practices, change and transformation are also observed in corporate communication processes. Along with factors such as changes in communication tools, rapidly increasing and difficult competition conditions, effective communication processes that need to be maintained with target audience and stakeholders, institutional reflections of digitalization, efforts to establish long-term and solid bonds, institutions are also trying to create an effective, two-way and engagement. Within the scope of these conditions, social media influencers also appear as an effective communication tool and element. It is the ideal and expected two-way and effective, continuous communication in the public relations process in which a person or institution is involved. The interaction factor is also important in the provision and continuity of this process. The engagement element takes place in an effective public relations process, mutual trust, information exchange, persuasion and sustainability. It can be said that social media influencers are positioned as a communication tool and element in order to ensure engagement.

The effectiveness of social media tools and environments in these communication processes cannot be denied. In the communication processes that the institutions plan and maintain with their target audience and stakeholders, a structure that progresses in two directions and where the presence of engagement is dominant is observed. This two-way communication process also overlaps with the interaction factor in the relevant tools and environments. While social media influencers can produce their own content in this process, they can also collaborate with relevant institutions. With these collaborations, information flow from the institution can be ensured and an effective public relations process can be carried out in addition to engagement with the followers.

As a field of activity, the Aviation industry includes the communication process where it is necessary and effective to establish an effective relationship with the target audience/customers. Therefore, they use effective communication tools to establish a long-term communication process with their customers, to establish a bond, to be preferred, to provide services to their customers. At this point, social media tools and environments are preferred and used effectively in this industry with their features such as effectiveness and interaction. In this communication environment, social media influencers, who are shining their stars day by day, draw attention. It emerges as an effective communication element at points such as creating interaction, making communication sustainable, bringing the corporate messages together with the target audience effectively, creating and receiving feedback.

As in different fields of activity, social media influencers are also seen in aviation industry. Aviation can be described as social media influencers. These people, who are described as social media influencers, are virtual influencers (non-real people), people/virtual people who are active in producing content related to any brand, product, service, idea, presenting existing content, bringing it together with the target audience, influencing and transforming it into behavior defined as persons. From this point of view, the effectiveness of the influencers in the sector, which are also described as aviation influencers, cannot be denied.

Influencers in the aviation industry, which includes many fields, are in different categories in their own fields. Some of these categories include areas such as travel, aviation, accommodation. Social media influencers in each field can share their experiences, views and opinions on the relevant field and produce content for them. In this direction, they can establish an effective and two-way communication with both the followers or the targeted audience, and the followers can also shape their attitudes and behaviors from time to time, and gain knowledge with these experiences and opinions. At this point, social media influencers in general and of course aviation industry social media influencers take place as an important and effective communication tool, element in the communication process of institutions with their target audience/followers. They have important and effective role in this process.

## REFERENCES

Катерна, О. К., & Молчанова, К. М. (2020). Digital Transformation of Ransformation of Aviation Industry in Ukraine. *ЕКОНОМІКА І УПРАВЛІННЯ*, 53-63.

Airport-Technology.com. (2021, February 9). *Biggest influencers in Airlines in Q4 2020: Top individuals and companies to follow.* Airport Technology. https://www.airport-technology.com/dashboards/influencers-dashboards/biggest-influencers-airlines/

Alam, M. A. (2016). Techno-stress and productivity: Survey evidence from the aviation industry. *Journal of Air Transport Management*, *50*, 62–70. doi:10.1016/j.jairtraman.2015.10.003

Allianz.com.tr. (2020). *Günümüzün Popüler Mesleği Influencer Nedir? Influencer Nasıl Olunur?* Allianz. https://www.allianz.com.tr/tr_TR/seninle-guzel/gunumuzun-populer-meslegi-influencer-nedir.html

Alraies, M. C., & Sahni, S. (2017). Why cardiologists should be on social media–the value of online engagement. *Expert Review of Cardiovascular Therapy*, *15*(12), 889–890. doi:10.1080/14779072.2017.1408408 PMID:29164942

Avram, B. (2017). Ancillaries in the aviation industry. Importance, trends, going digital. *Expert Journal of Marketing*, *5*(2), 53–65.

Barbe, D., & Neuburger, L. (2021). Generation Z and Digital Influencers in the Tourism Industry. In N. Stylos, R. Rahimi, B. Okumus, & S. Williams (Eds.), *Generation Z Marketing and Management in Tourism and Hospitality* (pp. 167–192). Palgrave Macmillan. doi:10.1007/978-3-030-70695-1_7

Bigne, E., Andreu, L., Hernandez, B., & Ruiz, C. (2018). The impact of social media and offline influences on consumer behaviour. An analysis of the low-cost airline industry. *Current Issues in Tourism*, *21*(9), 1014–1032. doi:10.1080/13683500.2015.1126236

Binthiyaa, B., & Fazıl, A. (2023). Impact of social media in airline industry. *Eur. Chem. Bull. 12*(4), 9141-9148. https://www.eurchembull.com/uploads/paper/80a0722b8ddad491b0c6d6550557eeba.pdf

Brodie, R. J., Hollebeek, L. D., Jurić, B., & Ilić, A. (2011). Customer engagement: Conceptual domain, fundamental propositions, and implications for research. *Journal of Service Research, 14*(3), 252–271. doi:10.1177/1094670511411703

Broekel, T., & Boschma, R. (2012). Knowledge networks in the Dutch aviation industry: The proximity paradox. *Journal of Economic Geography, 12*(2), 409–433. doi:10.1093/jeg/lbr010

Campbell, C., & Farrell, J. R. (2020). More than meets the eye: The functional components underlying influencer marketing. *Business Horizons, 63*(4), 469–479. doi:10.1016/j.bushor.2020.03.003

Chumwatana, T., & Chuaychoo, I. (2017, October). Using social media listening technique for monitoring people's mentions from social media: A case study of Thai airline industry. In *2017 2nd International Conference on Knowledge Engineering and Applications (ICKEA)* (pp. 103-106). IEEE. 10.1109/ICKEA.2017.8169910

Conde, R., & Casais, B. (2023). Micro, macro and mega-influencers on instagram: The power of persuasion via the parasocial relationship. *Journal of Business Research, 158*, 113708. doi:10.1016/j.jbusres.2023.113708

Dessart, L. (2017). Social media engagement: A model of antecedents and relational outcomes. *Journal of Marketing Management, 33*(5-6), 375–399. doi:10.1080/0267257X.2017.1302975

Dessart, L., Veloutsou, C., & Morgan-Thomas, A. (2015). Consumer engagement in online brand communities: A social media perspective. *Journal of Product and Brand Management, 24*(1), 28–42. doi:10.1108/JPBM-06-2014-0635

Digitalage.com.tr. (2023, March 9). *Markaların Influencer Marketing Yatırımlarının En Büyük Engeli Samimiyet.*. Digital Age. https://digitalage.com.tr/markalarin-influencer-marketing-yatirimlarinin-en-buyuk-engeli-samimiyet/

Dijkmans, C., Kerkhof, P., & Beukeboom, C. J. (2015). A stage to engage: Social media use and corporate reputation. *Tourism Management, 47*, 58–67. doi:10.1016/j.tourman.2014.09.005

Dožić, S. (2019). Multi-criteria decision making methods: Application in the aviation industry. *Journal of Air Transport Management, 79*, 101683. doi:10.1016/j.jairtraman.2019.101683

Dube, K., Nhamo, G., & Chikodzi, D. (2021). COVID-19 pandemic and prospects for recovery of the global aviation industry. *Journal of Air TransportManagement*, *92*, 102022. doi:10.1016/j.jairtraman.2021.102022 PMID:36567961

Ellrich, F., Bauer, M., Schreiner, N., Keil, A., Pfeiffer, T., Klier, J., Weber, S., Jonuscheit, J., Friederich, F., & Molter, D. (2020). Terahertz quality inspection for automotive and aviation industries. *Journal of Infrared, Millimeter and Terahertz Waves*, *41*(4), 470–489. doi:10.1007/s10762-019-00639-4

Fazıl, A. (2023). Impact Of Socıal Medıa In Aırlıne Industry. *Eur. Chem. Bull. 12*(4), 9141-9148.

Feedspot.com. (2023, August 7). *Top 45 Aviation Influencers in 2023.* FeedSpot. https://blog.feedspot.com/aviation_instagram_influencers/

Gill, G. K., & Shergill, G. S. (2004). Perceptions of safety management and safety culture in the aviation industry in New Zealand. *Journal of Air Transport Management*, *10*(4), 233–239. doi:10.1016/j.jairtraman.2004.02.002

Gisario, A., Kazarian, M., Martina, F., & Mehrpouya, M. (2019). Metal additive manufacturing in the commercial aviation industry: A review. *Journal of Manufacturing Systems*, *53*, 124–149. doi:10.1016/j.jmsy.2019.08.005

Hays, S., Page, S. J., & Buhalis, D. (2013). Social media as a destination marketing tool: Its use by national tourism organisations. *Current Issues in Tourism*, *16*(3), 211–239. doi:10.1080/13683500.2012.662215

Ibrion, M., Paltrinieri, N., & Nejad, A. R. (2019, October). On risk of digital twin implementation in marine industry: Learning from aviation industry. []. IOP Publishing.]. *Journal of Physics: Conference Series*, *1*(1), 012009. doi:10.1088/1742-6596/1357/1/012009

Influencermarketing.ai. (2023). *Top Travel, Tourism, And Aviation Instagram Influencers To Follow In Celle For 2023.* Influence Marketing. https://influencermarketing.ai/top-travel-tourism-and-aviation-instagram-influencers-to-follow-in-celle-for-2023/

Kang, M., & Schuett, M. A. (2013). Determinants of sharing travel experiences in social media. *Journal of Travel & Tourism Marketing*, *30*(1-2), 93–107. doi:10.1080/10548408.2013.751237

Karaman, A. S., Kilic, M., & Uyar, A. (2018). Sustainability reporting in the aviation industry: Worldwide evidence. *Sustainability Accounting. Management and Policy Journal*, *9*(4), 362–391.

Kim, B., Yoo, M., & Yang, W. (2020). Online engagement among restaurant customers: The importance of enhancing flow for social media users. *Journal of Hospitality & Tourism Research (Washington, D.C.)*, *44*(2), 252–277. doi:10.1177/1096348019887202

Knoblich, S., Martin, A., Nash, R., & Stansbie, P. (2017). Keys to success in Social Media Marketing (SMM)–Prospects for the German airline industry. *Tourism and Hospitality Research*, *17*(2), 147–164. doi:10.1177/1467358415610415

Linz, M. (2012). Scenarios for the aviation industry: A Delphi-based analysis for 2025. *Journal of Air Transport Management*, *22*, 28–35. doi:10.1016/j.jairtraman.2012.01.006

Malighetti, P., Meoli, M., Paleari, S., & Redondi, R. (2011). Value determinants in the aviation industry. *Transportation Research Part E, Logistics and Transportation Review*, *47*(3), 359–370. doi:10.1016/j.tre.2010.11.007

Marketingturkiye.com.tr. (2021, October 9). *Influencer Büyüsünü Güçlendirmek.* Marketing Turkiye. https://www.marketingturkiye.com.tr/haberler/influencer-buyusu/

Marketingturkiye.com.tr. (2023, March 26). *Sosyal Medyanın En Etkili Influencer'ları.* Marketing Turkiye.

Paek, H. J., Hove, T., Jung, Y., & Cole, R. T. (2013). Engagement across three social media platforms: An exploratory study of a cause-related PR campaign. *Public Relations Review*, *39*(5), 526–533. doi:10.1016/j.pubrev.2013.09.013

Pop, R. A., Săplăcan, Z., Dabija, D. C., & Alt, M. A. (2022). The impact of social media influencers on travel decisions: The role of trust in consumer decision journey. *Current Issues in Tourism*, *25*(5), 823–843. doi:10.1080/13683500.2021.1895729

Punel, A., & Ermagun, A. (2018). Using Twitter network to detect market segments in the airline industry. *Journal of Air Transport Management*, *73*, 67–76. doi:10.1016/j.jairtraman.2018.08.004

Scott, B. I., & Trimarchi, A. (2017). The Digital Aviation Industry: A Balancing Act Between Cybersecurity and European Consumer Protection. *Air and Space Law*, *42*(4/5), 443–462.

Smith, B. G., & Gallicano, T. D. (2015). Terms of engagement: Analyzing public engagement with organizations through social media. *Computers in Human Behavior*, *53*, 82–90. doi:10.1016/j.chb.2015.05.060

Smith, E., W., Kimbu, A. N., de Jong, A., & Cohen, S. (2023). Gendered Instagram representations in the aviation industry. *Journal of Sustainable Tourism*, *31*(3), 639–663. doi:10.1080/09669582.2021.1932933

Tian, X., He, W., Tang, C., Li, L., Xu, H., & Selover, D. (2020). A new approach of social media analytics to predict service quality: Evidence from the airline industry. *Journal of Enterprise Information Management*, *33*(1), 51–70. doi:10.1108/JEIM-03-2019-0086

Tikhonov, A. I., Sazonov, A. A., & Novikov, S. V. (2019). Digital aviation industry in Russia. *Russian Engineering Research*, *39*(4), 349–353. doi:10.3103/S1068798X19040178

Wagner, S. M., & Walton, R. O. (2016). Additive manufacturing's impact and future in the aviation industry. *Production Planning and Control*, *27*(13), 1124–1130. doi:10.1080/09537287.2016.1199824

Wang, X., & Liu, Z. (2019). Online engagement in social media: A cross-cultural comparison. *Computers in Human Behavior*, *97*, 137–150. doi:10.1016/j.chb.2019.03.014

Wardani, P. E., Marroy, D. S. H., Octora, Y., & Setiawan, E. B. (2019). Impact of Social Media Influencer Towards Brand Attitude and Purchase Intention in Airline Industry. *Advances in Transportation and Logistics Research*, *2*, 496–503.

Weckman, G. R., Shell, R. L., & Marvel, J. H. (2001). Modeling the reliability of repairable systems in the aviation industry. *Computers & Industrial Engineering*, *40*(1-2), 51–63. doi:10.1016/S0360-8352(00)00063-2

Wibisono, A. I., & Ruldeviyani, Y. (2021, April). Detecting Social Media Influencers of Airline Services through Social Network Analysis on Twitter: A Case Study of the Indonesian Airline Industry. In *2021 3rd East Indonesia Conference on Computer and Information Technology (EIConCIT)* (pp. 314-319). IEEE. 10.1109/EIConCIT50028.2021.9431876

Wong, K., Rudd, C., Pickering, S., & Liu, X. (2017). Composites recycling solutions for the aviation industry. *Science China. Technological Sciences*, *60*(9), 1291–1300. doi:10.1007/s11431-016-9028-7

Xiong, M., & Wang, H. (2022). Digital twin applications in aviation industry: A review. *International Journal of Advanced Manufacturing Technology*, *121*(9-10), 5677–5692. doi:10.1007/s00170-022-09717-9

# Chapter 5
# The Pros and Cons of Digitalization in Aviation:
## Navigating Digital Potential Risks Into Opportunities

**Tugba Erhan**
https://orcid.org/0000-0002-5697-490X
*Suleyman Demirel University, Turkey*

**Inan Eryilmaz**
https://orcid.org/0000-0001-8307-2402
*Suleyman Demirel University, Turkey*

## ABSTRACT

*The latest technology, digital implementations, and transformations have mainly shaped the development of the aviation industry. In addition to these developments, unprecedented risks must be considered to overcome the paradox of the beneficial and adverse sides of the new technology. The latest digital trends are explored and discussed by recognizing the importance of digital transformation in shaping the future of aviation. Thus, its role as a catalyst for innovation in the aviation industry, IATA (International Air Transport Association) is one of the pivotal organizations established to promote safe, reliable, and efficient air travel. However, in a digitalized world, the evolving nature of digital transformation within the latest technological developments and the power of technology effects are critical to take part at the forefront of the aviation industry. To provide insights into digital transformation in aviation management by discussing and evaluating the potential benefits and challenges, this chapter aims to contribute novel perspectives to digital transformation.*

DOI: 10.4018/979-8-3693-0732-8.ch005

*"If I had asked people what they wanted, they would have said faster horses."*

Henry Ford

## 1. INTRODUCTION

As Peter Drucker says, "One cannot manage change. One can only be ahead of it". Drucker's quote implies that change is inevitable, and as it cannot be controlled, it is not also possible to manage the change without being a part of and ahead of it. The word "change" also emphasizes establishing the future and shaping its impacts positively rather than being shaped by the forces of change. Organizations are the driving force and the source of the change. In addition to this potential trigger, technological innovations are added as one of the essential sources of a competitive business environment (Kok & Siripipatthanakul, 2023). Latest technological developments such as artificial intelligence (AI), robotics, and machine learning (ML) have been improving people's social and business environments in the era of the fourth industrial revolution (I4.0.) (Malik et al., 2021). The common goal of these revolutions is to enhance efficiency; new technologies also include complicated processes and unpredictable results (Woods and Dekker, 2000). The aviation industry is one of these which encounters the challenges (Xiong and Wang, 2022) of the digital transformation as well as its benefits, such as the Bluetooth-equipped unit load device (ULD) developed by the firm Unilode Aviation Solutions (Brett, 2019) or the digital twinning (DT) which provides significant opportunities for the aviation industry representing the virtual replicas of physical devices or products (Grieves, 2015). These digitally applied implementations are considered to improve and prioritize the safety of the overall operational process in the aviation industry by emphasizing the importance of digital safety for all aviation operations, referred to as general aviation (Hook et al., 2022). Digital technologies in the aviation industry help not only by the decrease in accident rates (Lu et al., 2006) but also by being the facilitator of the application of digitalization in aviation, airline organizations, employees, and customers may move beyond the actual applications for the sustainability (Lanshina et al., 2020).

The aviation industry has gone through many stages under the impacts of historical events since World War I (Reis, 2022). The latest technology, digital implementations, and transformations have mainly shaped the development of the aviation industry. In addition to these developments, the unprecedented risks (Ibrion et al., 2019) must be considered to overcome the paradox of the beneficial and adverse sides of the new technology, and transformation in the aviation industry (Javornik et al., 2019). Besides, launched in 2019, Accelerate@IATA presents an innovation program to accelerate innovation in the air transport industry. The International Air Transport Association creates an environment where the latest digital trends are explored and discussed (IATA, 2023) by recognizing the importance of digital transformation in shaping the future of aviation. Thus, its role as a catalyst for innovation in the aviation industry, IATA is a pivotal organization established to promote safe, reliable, and efficient air travel. Through its ongoing efforts, IATA fosters collaboration among industry stakeholders and remains dedicated to enhancing the aviation landscape to benefit passengers and the industry. However, in a digitalized world, the evolving nature of digital transformation within the latest technological developments and the power of technology effects are critical to take part at the forefront of the aviation industry. Also, airlines and aviation-related businesses rely on digital innovations to optimize operations and address sustainability challenges, supporting its members and partners in navigating this transformative process. To provide insights into digital transformation in aviation management by discussing and evaluating the potential benefits and challenges, this study aims to contribute novel perspectives to digital transformation.

## 2. BACKGROUND

### 2.1 Digitalization in the Aviation Industry

In the intense era of technology, both individuals and organizations target to achieve and adopt the latest version of the digital movement and become part of the digitalization process. Industries are mostly aware of these improvements by seeking ways to pursue and reach out to reveal their best potential, as digitalization has been one of the specific indicators of competitiveness (Jansen and Jeschke, 2018). The adoption of digital technologies has been reshaping industries' business models by interacting profoundly with the latest developments. In addition, to go beyond the digital era or integrate the new business models, organizations have been reshaping basic approaches, combining the stakeholder values with operating models during the current digitalization process to move forward with more digitalized principles (Berman, 2012). The aviation industry is one of these that has an essential role in facilitating global transportation for people and goods, supporting the tourism industry by connecting the local and international boundaries. Providing a vital service for humanity, the aviation industry is required to follow the latest technology as it is a highly technologically based industry (Frenken, 2000). Millions of passengers rely on the aviation sector daily to reach their destinations efficiently and safely. Moreover, the aviation industry provides essential contributions to the global economy by creating employment, supporting initiatives, pursuing technological developments, and using innovative approaches from aircraft engineering to air traffic management. Therefore, it can be highlighted that the aviation industry has more responsibilities far beyond flight operations.

Along with the profound advantages, the digital transformation also unveils a complicated series of challenges. Digitalization offers many opportunities for the aviation industry to enhance every aspect of its departments, including practical and immediate solutions. However, this transformation may face with challenging adaptations, and the workforce can be one of these in the adaptability process. Digitalized transformations require proactive participation by applying and adopting a holistic approach. For the last decade, undoubtedly, the crisis has been perceived as the Covid-19 pandemic, which caused tremendous changes at the individual and organizational levels. The aviation industry is one of the industries that has been affected severely. Travel restrictions, lockdowns, and a decrease in passenger demand led to a substantial reduction in flights, causing financial strains for airlines, airport operators, and related businesses. The aviation industry has been required to adapt rapidly by implementing safety protocols, exploring innovative sanitation measures, and reevaluating business strategies to navigate through these unprecedented crisis times. Besides, unforeseen challenges have affected the future of the aviation industry through these aforementioned agile responses being an essential part of the digital transformation process. Not only surviving throughout the unexpected crisis but also competing with the airline companies have been prioritized within the nonstop developments in the digitalized world that are aimed to be applied in all industries. Flight planning, arriving on time, or checking-in procedures take time and cause stress, especially for elderly passengers that result in travel stress (Petrenko, 2015) or travel anxiety that can be prevented or reduced with the help of the service quality and flight crew's competence (Batoueia et al., 2019). According to the study conducted by Westwood et al. (2000), passengers' safety and concern were found to be the primary factors that affected choosing the airline companies (Westwood, Pritchard, and Morgan 2000). In order to keep the staff well informed and update them with the newly changed digital applications, regular training should be planned to make them aware of how the potential challenges can turn into effective results that can help them perform their tasks and provide services. As a result of the investment in digital developments, it improves the service quality, digital ticketing, online check-in, and real-time tracking that make the passengers experience a comfortable and reliable flight time.

On the other hand, it should also be considered that when check-in counters are closed, this can also cause the loss of job opportunities. Not only passengers but also all employees can benefit from a digitally developed airline company. Plus, digitalization, in other words, digital transformation in the aviation industry, provides many advantages as well as disadvantages. However, the reliability of technological advancements without comprehending their implementations or evaluating the outcomes may lead to critically essential operational problems. Cybersecurity threats become more evident as aviation transforms into a digital mode, which keeps the passenger safety and privacy information of the passengers and the companies. Besides, pursuing the latest digital systems can also be expensive, specifically for smaller airline companies. While leading and bigger companies can easily afford and adapt the new digital tools or applications to their systems immediately, other smaller airline companies struggle to follow and implement the digital tools.

## 2.2 Aviation Employees' Approaches Towards Digitalization

The aviation industry has been facing threats and risks as well as opportunities throughout history. Specifically, the recent pandemic was one of the worst events that caused critical amounts of threats and sustainability risks for the aviation industry. According to some experts, it would require at least an additional six years for the airline industry to return to its usual levels of recovery (Belhadi et al., 2021).

Digital transformation in the aviation industry receives various reactions from aviation employees. Their approaches towards digitalization can be influenced by various factors, such as their role in the industry, previous experiences, and adaptability to change. Along with the positive reactions or perceptions, most probably the worst risk of all is the damage given to the needs of the human workforce (Parker and Grote, 2022). Employees are underpaid or lose their jobs (Selenko, 2022) when their tasks are replaced by digital technologies and this replacement can have severe effects on their well-being and mental health (Coldwell, 2019). Reactions towards digitalization may differ depending on the generation. Some are digital natives with tablets and smartphones (Prensky, 2001;), and some employees are uncomfortable with technology (Colbert et al., 2016). Being a significant part of the digital transformation (Van Der Schaft et al., 2022), human workforce learning abilities towards technological changes may vary; some employees are willing to use and adapt, while others experience challenges to integrate (Linnes and Metcalf, 2017). These employees comprehend the instant support and urgent need for digital transformation to embrace the constant changes. In addition, to approach closely to the new technological changes, digitally enhanced technologies encompass personal lives that may also emerge the dark side of digitalization referred to as opposing sides of the Digitalization of the Individual (DOTI) (Turel et al., 2019).

Finally, it is crucial to emphasize that digital transformation can make tasks easier, facilitate challenging jobs, improve personal and company safety, and support employees with the potential for further career opportunities by using the digitalization process. Besides, understanding the varied approaches of aviation employees towards digitalization is crucial for a smoother transition. Companies should pursue inclusive strategies both for digital natives and employees who experience social isolation due to digitalization, they should address and pay attention to individual concerns, revealing the strengths of each group that can help in effectively navigating the digital transformation of the aviation industry.

## 2.3 Sustainability of the New Implementations in the Aviation Industry

Sustainability has become a focal point in many industries. The aviation industry is also one of these industries that targets to consider and evaluate the environmental concerns and reduce its carbon footprint greenhouse gas emissions as it is one of the significant industries increasing the carbon level in order to assess and attract attention to decarbonize aviation, IATA's World Sustainability Symposium will be held in Spain to look for solutions to this biggest challenge in the aviation industry. The symposium aims to achieve net zero emissions by 2050. (IATA, 2023). Starting from airline companies to aircraft manufacturers and all aviation industry stakeholders come to terms with the fact that sustainability in aviation can be risky unless decarbonization is achieved to obtain more environmentally friendly policies. However, when airline companies invest in new technology, they aim to decrease operational costs as well as reduce emissions (Bhatti et al., 2021). For the aviation industry, implementing the new technology has been forced to adapt to be sustainable. According to ATAG's data presented in 2019 (Airport Transport Action Group), every day, 4.5 billion passengers are being transferred, and 48.8 million flights are being performed.

In 2019, the aviation industry worldwide contributed 914 million tons of CO2 emissions, while on a global scale, human activities generated more than 43 billion tons of CO2 emissions (ATAG, 2019). The other industries achieve a certain amount of reduction in greenhouse gas emissions (GHG); the aviation industry is the most critical industry to decarbonize (Chiaramonti, 2018). Sustainable aviation fuels (SAFs) share a chemical composition similar to traditional jet fuel. Yet, they are derived from renewable resources, offering the potential to lower the overall emissions throughout their lifecycle (Gonzalez-Garay et al., 2022). Most of all, it is essential to measure these parameters to track them efficiently for sustainability through modern and non-traditional data sources (Fritz et al., 2019). Besides, according to the United Nations Sustainable Development Goals (UN SDGs) industries have been rushing to achieve their best potential to provide a contribution to a sustainable environment by the year 2030, however, the time is short, and the industries may be late to prioritize the sustainability goals along with other goals (Sankaran et al., 2020). The aviation industry has also undertaken several tasks starting from the reduction of Greenhouse Gas emissions (Elhmoud and Kutty, 2020). The fuel use and greenhouse gas emissions of the aviation industry increased dramatically over the years, causing irreversible adverse effects on the environment and human survival (Dessens et al., 2014) due to the increasing demand for tourism and cargo deliveries (Dube et al., 2021). Thus, the sustainability and continuity of the reciprocal process for humanity, the aviation industry becomes a part of the green aviation industry, also known as eco-friendly aviation (Rice et al., 2020).

By proactively addressing the challenges and navigating through potential risks, the aviation industry can transform these into opportunities. This not only provides a safe and efficient environment but also results in a more connected, sustainable, and passenger-centric future in aviation. In other words, the evolution of the aviation industry in the digital age will depend on its resilience and adaptability. The aviation industry moves towards sustainability, encompassing technological innovations, operational changes, and shifts in management strategies. Furthermore, sustainability goals, when adapted as strict principles, support the companies during unexpected or unplanned crisis periods (Agyabeng-Mensah et al., 2020).

## 3. SOLUTIONS AND RECOMMENDATIONS

Digital transformation in the aviation industry has the potential to significantly enhance sustainability by improving operational efficiency, reducing fuel consumption, and minimizing environmental impact. It also enhances the passenger experience and offers valuable data for monitoring and optimizing sustainability efforts. However, the industry must address the challenges and risks associated with digital transformation to fully realize its sustainability benefits. Newly developed wireless technologies can be applied that foster connectivity or several entertainment activities can be provided in order to strengthen the bond between human-machine collaboration during the flights. Finally, it is more evident that the price of cost, increase in inflation rate, and fluctuation in oil prices cause challenges. The aviation industry needs to adopt the benefits of technologies such as electric or hybrid aircraft models. Optimizing flight paths and operations, reducing fuel consumption, enhancing passenger safety, protecting personal data and flight safety, and using biometric check-ins are just some of the benefits of the digital transformation in the aviation industry. However, following all processes can be supported by advanced data analysis and up-to-date advancements in technology. Continuous feedback from internal and external stakeholders can also provide contribution through digital services gathered from innovative digital mechanisms. In addition, digital systems should be used to obtain real-time data such as the weather conditions or the accuracy of flight safety. Aircraft staff training should be held regularly in order to share the latest digital tools and newly applied flight protocols. Like all industries, the aviation industry will also gain long-term cost savings by applying or benefiting from the latest digitalized movements. Aviation companies that lead as strong competitors in sustainability should be involved in more digital interventions presented in the field of aviation. One of the most important advantages is the reduction of the carbon footprint by performing efficient routes and operations. Therefore, the aviation industry can reduce the industry's carbon footprint by following and integrating the environmental metrics. Starting from the passengers, all other stakeholders of the aviation industry prefer to fly by an airline company that can guarantee the protection of privacy, so by implementing digital solutions, internal and external stakeholders can prioritize and choose the best company that provides instant digital solutions. However, integrating new digital systems into existing infrastructure can be complex. Specialists and consultants should be employed during the integration of the digital change or update process. As digital developments bring automation that may reduce the need for more workforce. The human workforce should be planned carefully; otherwise, digital systems can only survive on robotic technology when digitally skilled employees face losing their jobs. Employees can have training to adapt and manage the system constantly. While digital transformation in the aviation industry offers advantages, it's essential to address the inevitable challenges proactively.

## 4. FUTURE RESEARCH DIRECTIONS

The current study provides contributions to future research in aviation management in the digital age. During the digital era, several developments have been added, such as artificial intelligence (AI) and machine learning. As a result of this deep integration into the aviation industry, sustainability, passenger experiences, and operational competencies have been improved. Within these objectives, IATA has recognized the significance of digital transformation in shaping the future of aviation. In accordance with the developments in AI, air traffic control, ticketing process, identity verification, and supply chain can be regulated for future developments in the aviation industry. This book chapter emphasizes the role of IATA and its importance; besides, digital transformation is considered to provide an increasingly safety service for cybersecurity. Also, it is an inevitable fact that as aviation becomes more digital, the risks of cyber threats increase. Thus, it is recommended that the focus of digital transformation should aim to be aware of the evolving threats and the solutions to mitigate these potential threats while benefiting the advantages to compete and become the leading industry among the others. These threats can be perceived as vulnerable parts of aviation security against cyberattacks on air traffic control systems; however, as mentioned earlier in the chapter, digital twin technology (Bhatti et al., 2021) can be improved by using virtual environments of physical methods to predict and optimize operational efficiencies. In addition, virtual simulations play an essential role in the pilot's training, assisting the ground crew to perform the tasks even in unpredicted emergencies that require instant intervention. Another aspect of digital transformation efforts also supports and enhances the sustainability goals in the aviation industry. In order to allocate resources efficiently to reduce waste, digital tools will be improved more to optimize overconsumption. This rapid pace in technology evolves and adds continuous changes, mostly in a positive way, to promote a safer and more efficient environment for both the employees and passengers in the aviation industry. Also, as the aviation industry provides services for passengers, it has an essential role in global transportation, connecting people and goods across the world. Therefore, adopting the new and latest technological developments, the aviation industry is required to explore each aspect of implications, including and encompassing the ethical safety for the passengers, prioritizing their comfort to improve the overall travel experience constantly during the digital transformation period. Besides, this is the era of digital tools prevalence, so future research may look for ways to establish the best connection and adaptation between humans and machines in the digitally evolved aviation industry.

## 5. CONCLUSION

The inevitable changes in digitalization offer opportunities in the aviation industry. However, the digital transformation in aviation management engages with potential benefits and risks that can be perceived as challenges. Like other disciplines, the aviation industry is one of the most dynamic sectors where embracing digitalization becomes a competitive advantage and a necessity for long-term sustainability. As aviation stakeholders harness the power of data analytics, automation, and connectivity, they not only enhance operational efficiency and passenger experiences but also face the critical task of ensuring robust cybersecurity measures to protect against evolving threats in an increasingly interconnected world. Starting from the safety strategies, potential unprecedented crises may cause severe consequences in the aviation industry. Traditional safety strategies may have been effective in maintaining a high level of safety; however, more is needed within the evolving aviation industry. Thus, to enhance the evolution surrounded by the latest technology, digital transformation can be perceived as a complementary tool to traditional aviation strategies. Fostering by being a part of newly added technological implementations, the aviation industry keeps aiming for new competencies in innovation. Integrating emerging technologies like artificial intelligence or machine learning offers opportunities for reaching the maximum prediction levels on real-time decision-making processes during flight operations, ground services, air traffic management, or supply chains. However, it is possible and related to having a sense of vision to keep up with these developments, but problems may occur in adopting and integrating these technologies without giving any harm to safety and reliability. Moreover, the human workforce within the aviation ecosystem cannot be underestimated by the complexity of digitalization. The industry must invest in continuous training and skill development strategies considering that the workforce will not only be adapting to using these new tools but also must be ready to comprehend their implications. As a result, human-based or human-centered decisions are made by using data-driven and technology-oriented approaches.

In conclusion, as the aviation industry maintains its digital transformation journey, it is considered crucial to keep the balance between the benefits and potential risks of digital transformation which is required to be balanced. Adopting digitalization should not be perceived as the ultimate pursuit of technological development; in addition, digitally renewed operational capacities, safety protocols, and passenger experiences should be prioritized within this digital era. The aviation industry's success will be determined by setting new digitally competitive ways to set new benchmarks for the future goals of the aviation industry.

## 6. ACKNOWLEDGMENT

This research received no specific grant from any funding agency in the public, commercial, or not-for-profit sectors.

## REFERENCES

Agyabeng-Mensah, Y., Ahenkorah, E., Afum, E., Agyemang, A. N., Agnikpe, C., & Rogers, F. (2020). Examining the influence of internal green supply chain practices, green human resource management and supply chain environmental cooperation on firm performance. *Supply Chain Management*, *25*(5), 585–599. doi:10.1108/SCM-11-2019-0405

ATAG. (2020). *Facts*. Aviation Benefits Beyond Borders. https://atag.org/facts-figures/

Batouei, A., Iranmanesh, M., Nikbin, D., & Hyun, S. S. (2019). Flight anxiety: Investigating the role of airline service quality and flight crew's competence. *Asia Pacific Journal of Tourism Research*, *24*(7), 710–724. doi:10.1080/10941665.2019.1630457

Belhadi, A., Kamble, S., Jabbour, C. J. C., Gunasekaran, A., Ndubisi, N. O., & Venkatesh, M. (2021). Manufacturing and service supply chain resilience to the COVID-19 outbreak: Lessons learned from the automobile and airline industries. *Technological Forecasting and Social Change*, *163*, 120447. doi:10.1016/j.techfore.2020.120447 PMID:33518818

Berman, S. J. (2012). Digital transformation: Opportunities to create new business models. *Strategy and Leadership*, *40*(2), 16–24. doi:10.1108/10878571211209314

Bhatti, G., Mohan, H., & Singh, R. R. (2021). Towards the future of smart electric vehicles: Digital twin technology. *Renewable & Sustainable Energy Reviews*, *141*, 110801. doi:10.1016/j.rser.2021.110801

Brett, D. (2019) Unilode and OnAsset complete Bluetooth ULD trial. AirCargoNews. https://www.aircargonews.net/airlines/unilode-and-onasset-completebluetooth-uld-trial/

Chiaramonti, D. (2019). Sustainable aviation fuels: The challenge of decarbonization. *Energy Procedia*, *158*, 1202–1207. doi:10.1016/j.egypro.2019.01.308

Colbert, A., Yee, N., & George, G. (2016). The digital workforce and the workplace of the future. *Academy of Management Journal*, *59*(3), 731–739. doi:10.5465/amj.2016.4003

Coldwell, D. A. (2019). Negative influences of the 4th industrial revolution on the workplace: Towards a theoretical model of entropic citizen behavior in toxic organizations. *International Journal of Environmental Research and Public Health*, *16*(15), 2670. doi:10.3390/ijerph16152670 PMID:31349722

Dessens, O., Köhler, M. O., Rogers, H. L., Jones, R. L., & Pyle, J. A. (2014). Aviation and climate change. *Transport Policy*, *34*, 14–20. doi:10.1016/j.tranpol.2014.02.014

Dube, K., Nhamo, G., & Chikodzi, D. (2021). COVID-19 pandemic and prospects for recovery of the global aviation industry. *Journal of Air Transport Management*, *92*, 102022. doi:10.1016/j.jairtraman.2021.102022 PMID:36567961

Elhmoud, E. R., & Kutty, A. A. (2021). Sustainability assessment in aviation industry: A mini-review on the tools, models and methods of assessment. In *Proceedings of the 2nd African International Conference on Industrial Engineering and Operations Management, Harare, Zimbabwe* (pp. 7-10). IEOM Society. http://ieomsociety.org/harare2020/papers/324.pdf

Frenken, K. (2000). A complexity approach to innovation networks. The case of the aircraft industry (1909–1997). *Research Policy*, *29*(2), 257–272. doi:10.1016/S0048-7333(99)00064-5

Fritz, S., See, L., Carlson, T., Haklay, M., Oliver, J. L., Fraisl, D., Mondardini, R., Brocklehurst, M., Shanley, L. A., Schade, S., Wehn, U., Abrate, T., Anstee, J., Arnold, S., Billot, M., Campbell, J., Espey, J., Gold, M., Hager, G., & West, S. (2019). Citizen science and the United Nations sustainable development goals. *Nature Sustainability*, *2*(10), 922–930. doi:10.1038/s41893-019-0390-3

Gonzalez-Garay, A., Heuberger-Austin, C., Fu, X., Klokkenburg, M., Zhang, D., van der Made, A., & Shah, N. (2022). Unravelling the potential of sustainable aviation fuels to decarbonise the aviation sector. *Energy & Environmental Science*, *15*(8), 3291–3309. doi:10.1039/D1EE03437E

Grieves, M. (2015). *Digital twin: Manufacturing excellence through virtual factory replication*. (Digital twin white paper). FIT. https://research.fit.edu/media/sitespecific/researchfitedu/camid/documents/1411.0_Digital_Twin_White_ Paper_Dr_Grieves.pdf

Hook, L. R., Sizoo, D., & Fuller, J. G. (2022). *How digital safety systems could revolutionize aviation safety. In 2022 IEEE/AIAA 41st Digital Avionics Systems Conference (DASC)*. IEEE. https://www.iata.org/en/events/all/world-sustainability,

IATA. (2023). World Sustainability Symposium (WSS). IATA. https://www.iata.org/en/events/all/world-sustainability-symposium/

IATA. (2023). Digital Innovation Program. IATA. https://www.iata.org/en/programs/innovation/digital-innovation/

Ibrion, M., Paltrinieri, N., & Nejad, A. R. (2019). On risk of digital twin implementation in marine industry: Learning from aviation industry. *Journal of Physics*, *1357*(1), 012009.

Jansen, C., & Jeschke, S. (2018). Mitigating risks of digitalization through managed industrial security services. *AI & Society*, *33*(2), 163–173. doi:10.1007/s00146-018-0812-1

Javornik, M., Nadoh, N., & Lange, D. (2019). Data is the new oil. In B. Müller & G. Meyer (Eds.), *Towards User-Centric Transport in Europe. Lecture Notes in Mobility*. Springer. doi:10.1007/978-3-319-99756-8_19

Kok, S. L., & Siripipatthanakul, S. (2023). Change Management Model in Corporate Culture and Values: A Case Study of Intel Cooperation. *Advance Knowledge for Executives*, *2*(1), 1–30.

Lanshina, T. A., Barinova, V. A., Kondratyev, A. D., & Romantsov, M. V. (2020). Sustainable development and digitalization: The unusual COVID-19 crisis requires original solutions. *Bulletin of International Organizations*, *15*(4), 91–114. doi:10.17323/1996-7845-2020-04-05

Linnes, C., & Metcalf, B. (2017). iGeneration and their acceptance of technology. [IJMIS]. *International Journal of Management & Information Systems*, *21*(2), 11–26.

Lu, C. T., Wetmore, M., & Przetak, R. (2006). Another approach to enhance airline safety: Using management safety tools. *Journal of Air Transportation (Reston, Va.)*, *11*(2), 113–139.

Malik, N., Tripathi, S. N., Kar, A. K., & Gupta, S. (2021). Impact of artificial intelligence on employees working in industry 4.0 led organizations. *International Journal of Manpower*, *43*(2), 334–354. doi:10.1108/IJM-03-2021-0173

Parker, S. K., & Grote, G. (2022). Automation, algorithms, and beyond: Why work design matters more than ever in a digital world. *Applied Psychology*, *71*(4), 1171–1204. doi:10.1111/apps.12241

Petrenko, E. (2015). Designing effective pre-flight airport security information in electronic format: *Case study: Vilnius International Airport*. Theseus. https://www.theseus.fi/bitstream/handle/10024/101650/Petrenko_Ernestas2a.pdf?sequence=1

Prensky, M. (2001). Digital natives, digital immigrants. *On the Horizon*, *9*(5), 1–6. doi:10.1108/10748120110424816

Reis, M. F. (2022). *Evolution of the Aviation Industry: Impact of Digital Transformation of Learning Solutions.* HDL. http://hdl.handle.net/10400.6/13034

Rice, C., Ragbir, N. K., Rice, S., & Barcia, G. (2020). Willingness to pay for sustainable aviation depends on ticket price, greenhouse gas reductions and gender. *Technology in Society*, *60*, 101224. doi:10.1016/j.techsoc.2019.101224

Sankaran, S., Müller, R., & Drouin, N. (2020). Creating a 'sustainability sublime' to enable megaprojects to meet the United Nations sustainable development goals. *Systems Research and Behavioral Science*, *37*(5), 813–826. doi:10.1002/sres.2744

Selenko, E., Bankins, S., Shoss, M., Warburton, J., & Restubog, S. (2022). Artificial intelligence and the future of work: A functional-identity perspective. *Current Directions in Psychological Science*, *31*(3), 272–279. doi:10.1177/09637214221091823

Turel, O., Matt, C., Trenz, M., Cheung, C. M., D'Arcy, J., Qahri-Saremi, H., & Tarafdar, M. (2019). Panel report: The dark side of the digitization of the individual. *Internet Research*, *29*(2), 274–288. doi:10.1108/INTR-04-2019-541

Van Der Schaft, A. H., Lub, X. D., Van Der Heijden, B., & Solinger, O. N. (2022). How employees experience digital transformation: A dynamic and multi-layered sensemaking perspective. *Journal of Hospitality & Tourism Research (Washington, D.C.)*, 1–18. doi:10.1177/10963480221123098

Westwood, S., Pritchard, A., & Morgan, N. (2000). Gender-Blind Marketing: Businesswomen's Perceptions of Airline Services. *Tourism Management*, *21*(4), 353–362. doi:10.1016/S0261-5177(99)00069-2

Woods, D., & Dekker, S. (2000). Anticipating the effects of technological change: A new era of dynamics for human factors. *Theoretical Issues in Ergonomics Science*, *1*(3), 272–282. doi:10.1080/14639220110037452

Xiong, M., & Wang, H. (2022). Digital twin applications in aviation industry: A review. *International Journal of Advanced Manufacturing Technology*, *121*(9-10), 5677–5692. doi:10.1007/s00170-022-09717-9

# Chapter 6
# Evolution of Unmanned Aerial Systems and Inconsistencies Between Strategies, Concepts, and Technology

**Huseyin Onder Aldemir**
https://orcid.org/0000-0002-8083-0447
*Ozyegin University, Turkey*

## ABSTRACT

*Unmanned aerial systems (UAS) are undoubtedly a rising value that marked the 21st century in the aviation industry. Recently, both the high and precise technology used in aviation and the rapid development of artificial intelligence have enabled these remotely controlled aircrafts to find widespread use in the civilian field. Even these smart technologies and artificial intelligence transformed the UAS into different forms like flying swarm in the airspace and Urban Air Mobility (UAM) which comprises passenger and cargo transportation without pilot in metropolitan cities. In this chapter, the technological evolution of UAS is examined in the historical process, the components that make up UAS are explained, and UAS classification considering various characteristics with different aspects is detailed. Finally, issues resulting from strategies, concepts, and regulations falling behind evolving UAS technologies with aviation management perspective are discussed.*

DOI: 10.4018/979-8-3693-0732-8.ch006

# 1. INTRODUCTION

One of the greatest discoveries of the 20th century is undoubtedly unmanned systems. Studies in the fields of automation and artificial intelligence led to the development of unmanned systems. Unmanned Aerial Vehicles (UAVs) play a crucial role in the world of unmanned systems. UAVs are being studied in both the natural and social sciences as a result of their recent popularity and advancements.

Although UAVs describe an aircraft that flies without humans on board, a UAV is not an object that can fly on its own. In order for the UAV to fly, it requires a systematic approach and human-intensive operational work. This reveals the teamwork, task responsibilities and authorities of the people who will take part in the system. Ultimately, a team that can perform harmonious tasks is needed. Depending on the size of the UAVs and the nature of the operational work, different numbers of people must be included in the task teams of UAS. This number can range from one to tens of individuals. While a hobbyist or mini-class UAV used for any purpose may just require one person, larger UAV flight operations require a large team. Because of this, UAS operations necessitate a lot of coordination from humans even though it is referred to as an "Unmanned" system. The human factor sensitivity that applies to manned aircraft also applies to UAS.

Understanding and learning UAS are based on an interdisciplinary manner. UAS are utilized in a variety of contexts and advance daily in lockstep with changes in technology, strategy, and needs. The production phase of the aircraft, ground control station, and ground data terminal in these systems requires the contribution of many engineering fields, such as aerospace, electronics, mechanical, mechatronic, material, and industrial engineering. However, disciplines like strategy, business, management, law, and meteorology are also included in the application phase. Understanding UAS requires a synergistic system management approach. A separate research must be conducted to examine the technical challenge of choosing the location where UAS will be deployed and used.

The history of UAV means the history of aviation and dates back to before the discovery of manned aircraft. However, it failed to attract the attention of International Civil Aviation Organization (ICAO) until 2005 since UAV is not considered in an aircraft category. Therefore, UAS has not taken seriously enough until recently. ICAO did not even want to take responsibility for UAVs at first. Therefore, the efforts of aviation authorities worldwide to establish concepts and design strategies fall behind the quick and superior advancements in Unmanned Aerial System (UAS) technologies in recent years.

The aim of this chapter is:

- to show the technological evolution of the UAS throughout the history,
- to explain the components comprising the system, and working principles
- to elaborate the classification of UAS according to miscellaneous characteristics,
- to explain that the lack of interest of ICAO and other aviation authorities in UAS leads to incompatibility between the concept and technology,
- to illuminate the transformation of UAS into the UAM form,
- to clarify the emerging challenges during the transformation.

Following the introduction, a concise technological history of UAS is provided to help readers understand how UAS have incorporated all previous advancements in aviation and remote control technology. The components of UAS were examined under the heading of "Components of UAS" in order to realize that UAS is a system. In the following section, UAS classification is made in detail by considering different characteristics and the most accepted NATO UAS classification in the world is presented. Before the conclusion part, inconsistencies between strategies, concepts, and technology on UAS are discussed and arising challenges are demonstrated.

## 2. CONCISE HISTORY OF UAS TECHNOLOGICAL EVOLUTION

The place of Unmanned Aerial Vehicle Systems (UAS) in aviation history is a very comprehensive subject and a book could be written on significance of UAS. The purpose of this section is to give an insight about the development of UAV systems and to form a general perspective on the path followed by UAVs. When UAS are studied historically, it becomes clear that military requirements are what led to the development of non-human or remotely operated aircraft. The photograph, snapshot transmission and signal intelligence provided by UAVs have provided great opportunities for situational awareness, framing of the battlefields, maintaining, and concluding the operations for the military command staff. After manned aircraft were introduced, UAV development truly began with the earliest advancements in the fields of radio control and automation. Initial UAV development efforts emerged to serve the purpose of protecting Navy ships from warplanes in the US and UK. Concisely, wars and the needs of battlefields have driven the development of UAS.

In the modern sense, the emergence of UAV systems on the stage of history coincides with the times of the First World War (WW). Modern sense refers to presence of an automation system in aircraft. Although the need for UAVs emerged during the war, the tests and construction stages of two UAVs (Sperry-Curtiss and Kettering Bug) came to an end and were not employed in the First WW. In the previous periods, simple flying objects, kites or balloons were used for intelligence purposes such as reconnaissance, surveillance, image acquisition and target detection. The historical development of UAVs should be examined in 4 key stages:

- WW I period (1914-1918) and before,
- The period between two world wars (1919-1939),
- WWII period (1939-1945),
- Cold War (1947-1991) and Post-Cold War period.

## 2.1 World War I Period (1914-1918) and Before

The first collective use of unmanned aerial vehicles on the battlefield was an air attack by the Austrians in 1849 by bombing the city of Venice with 200 unmanned balloons loaded with time bombs. In 1898, Nikola Tesla discovered remote control with radio waves, patented this invention and laid the foundations of modern robotics (Monash University e-library, 2006). The emergence of UAVs in the modern sense is based on the application of remote control systems to aircraft after this date.

Tesla, an electrical genius, was busy designing the first electrical architecture of New York City, while another inventor, Elmer Sperry, founder of the famous flight control company that bears his name today, was developing the first gyrocontrol system. Sperry's work was initially focused on underwater torpedoes designed for the Navy, just like Tesla's. Sperry developed a three-axis mechanical gyroscope system that takes inputs from gyros and converts them into simple magnetic signals (Barhart et al., 2012).

In 1911, just 8 years after the start of manned flight, the inventor of the gyroscope, Elmer Sperry, became interested in the application of radio control to airplanes. With financial support and assistance from the US Navy, Sperry proved successful in applying the gyroscope to stabilized flight in 58 flight tests conducted by Lieutenant P. N. L. Bellinger at Hammondsport, New York, from August 31 to October 4, 1913 (Coletta, 1987). The team, consisting of Elmer Sperry, Peter Cooper Hewitt, Josephus Daniels and Glenn Curtiss, made the first UAV demonstration flight, an aerial torpedo, on behalf of the Navy on 21 November 1917. On March 6, 1918, the Sperry-Curtiss air torpedo made its longest successful flight of approximately 1,000 yards (914.4 meters). Thus, the first UAS flight was carried out. Experiments with pilotless flights continued, and the non-pilot N-9 aircraft, which was successfully launched on 17 October 1918, flew on its pre-arranged route, but failed during landing and crashed over the sea (Keane & Carr, 2013).

The Land Forces aircraft board requested Charles Kettering to create an unmanned "flying bomb" that could hit a target 50 miles away after learning about the US Naval Forces' efforts. In the project, Orville Wright was the consultant for the design of the aircraft and Childe H. Wills from Ford Motor Company was the consultant for the design of the engine, and Kettering's design was completed under the name "Kettering Bug" (Keane & Carr, 2013). The Kettering Bug aerial bomb performed impressively in tests, traveling a distance of 100 miles and reaching an altitude of 10,000 feet. To validate the airframe components, a model was built with a manned cockpit so that a test pilot could fly the aircraft. Unlike the Navy's decommissioned aerial torpedo, Kettering Bugs with aerial bombs were the first mass-produced unmanned aerial vehicle. Testing continued for approximately 12 to 18 months after the First World War (Barnhart et al, 2012). As can be seen, the first UAV development efforts in the modern sense are aerial torpedoes made by the USA.

## 2.2 The Period Between Two World Wars (1919-1939)

The majority of UAV development during and just before the First World War focused on building platforms for weapons like airborne torpedoes or bombs. In the post-World War I period, with the beginning of the researches to enhance the capabilities of the land and naval armed forces against the threats that might come from the air, steps had been taken for the development of target drones that can be air targets.

In the interwar period, the development of UAVs was influenced by other developments in the aviation world, especially in air transport. Unmanned target aircraft were needed due to requirements in the military field, even if advancements in air transportation had hampered the development of UAV systems. The Navy of the USA and England wanted to develop air defense weapon systems used at sea and train personnel as a result of their efforts to take precautions against warplanes in order to prevent their ships from being sunk by aircraft. At this point, target drones came on the scene.

The aviation industry has made ground-breaking discoveries, as in every other industry, but it is also evident that some technologies have evolved to advance the system's lower levels. Archibald Montgomery Low has developed data links, one of the most important inventions for UAV systems. Born in England in 1888, Professor Low is known as the "Father of Radio Guidance Systems". He developed the first data link and solved the interference problems caused by the UAV engine. In the first flight attempt, their UAV crashed, but on September 3, 1924, they made a successful radio-controlled flight (Fahlstrom & Gleason, 2012). While the British asserted that this was the first radio-controlled flight, the USA claimed on September 15, 1924, that the Curtiss F-5L had successfully completed all aspects of radio-controlled, long-distance flight, including takeoff, maneuver, and landing. Three more test flights were made on the same day and the tests continued for 14 months. The project was dormant until 1936 as a result of a failed test on December 11, 1925, which lowered interest in UAVs in the USA (Keane & Carr, 2013). The US Navy returned to the UAV arena in the late 1930s with the N2C-2 Target Aircraft developed by the Naval Research Laboratory. This 2500-pound (1134-kg) target aircraft has been quite effective in identifying the shortcomings of the Navy's air defense weapons (Barnhart et al., 2012).

In the interwar period, the British Royal Navy attempted to develop an unmanned aerial torpedo and an unmanned target aircraft, both using the same fuselage. This aircraft was attempted to be launched from the ship several times, but failed. As a result of the efforts of the Royal Aircraft Establishment and subsequently the Royal Air Force, the Fairey Scout 111 F, a manned aircraft, was transformed into a radio-controlled and gyrostable unmanned target aircraft and was named Queen. Five unmanned target planes were constructed for this purpose, but only one was used as a target in naval artillery fire. The highly stable DeHavilland Gypsy Moth aircraft was then equipped with the Fairey flight control system in the following step. Queen Bee was the name given to this unmanned target aircraft. The Royal Air Force has ordered 420 target aircraft Fairey Queen Bee. Later, the letter Q began to be used to denote unmanned operation. The letter Q was also used by the US Armed Forces to represent UAV operations. Although there is no confirmation, it is rumored that the word drone, which literally means drone, is derived from Queen Bee (Barnhart et al., 2012). Queen Bee is considered to be the ancestor of today's UAVs.

## 2.3 WWII Period (1939-1945)

During the Second World War, the target UAV and radio-controlled weapon UAVs were produced in large numbers and became the force multiplier of the battlefield throughout the war. Germany produced and employed guided missiles during the Second World War, joining the United States and England in UAV-related activities.

The U.S. Navy developed the TDN-1, a large-scale aerial torpedo, in the early 1940s, after the N2C-2 target UAV developed in the 1930s and controlled by another manned aircraft nearby. In order to limit crew risk, this aircraft is built to carry bombs or torpedoes in high-risk environments. The revolutionary development of this UAV was the installation of a primitive 75-pound (1 pound=0.4536 kg) RCA television camera in the nose of the aircraft as a detecting sensor to direct the pilot away but within a safe distance. Given the relatively poor reliability and resolution of the early TVs, this new technology integration was a remarkable achievement. TDN-1s were replaced by a more advanced model, Navy TDR-1 Assault UAVs. 140 units were prepared from these UAVs and the Special Air Task Force (SATFOR) was formed from them and sent to the Pacific battlefield. It was used against the Japanese at the Battle of Bougainville Island in 1944 and had limited success (Barnhart et al., 2012).

In 1937, another British, Reginald Leigh Denny, and two Americans, Walter Righter and Kenneth Case, developed a series of UAVs called the RP-1, RP-2, RP-3 and RP-4. They formed a company called the Radioplane Company in 1939 and later became part of the Northrop-Ventura Division. Radioplane produced thousands of target aircraft during the Second World War (Fahlstrom & Gleason, 2012). Denny drew attention in the demonstration he made at the request of the Artillery branch in 1938. A contract was signed by the US Army for the production of the RP-4, later designated OQ-1 (Blom, 2010). Initial purchases were few until the attack on Pearl Harbor on December 7, 1941. Subsequently, the Radioplane company produced more than 15,000 RP-5 radio-controlled target UAVs, making it the first mass-produced unmanned aircraft in history (Singer, 2009). These target UAVs were named OQ-2 by the US Army and TDD-1 (Target Drone Denny 1) by the Navy.

The OQ-2 is designed to attract anti-aircraft fire of enemy forces, to stall the air defense of the enemy and to increase the accuracy of friendly fire as a target. Operating with a 6 horsepower, 2-cylinder, 2-stroke engine, the OQ-2 can take off from the runway or launch from catapult, and land on the runway or with a parachute. It can stay in the air for 70 minutes at a maximum altitude of 8,000 ft. above ground level (AGL) and can reach a maximum speed of 85 miles per hour (Sloggett, 2015).

During World War II, Denny and his team made many changes to this UAV, increasing its capabilities and improving it to meet service needs. Denny has sold 50 OQ-1s, 5822 OQ-3s and 2084 OQ-4s to the Army and Navy. With technological developments and mechanical improvements, there has been a development from an OQ-2 powered by a 6 horsepower engine that can fly up to 85 miles per hour, to an OQ-14 with 14 horsepower and a flight speed of up to 140 miles per hour. UAV missions carried out during this period continued throughout the post-war period, until the OQ-19 was first used as a reconnaissance aircraft (Newcome, 2004).

The most important unmanned aircraft of the Second World War was undoubtedly the V-1 Buzz Bomb, which Nazi Germany called the Vengeance Weapon-1. The V-1 represents the first mass-produced cruise-missile type unmanned aircraft, and its configuration influenced many post-war drone designs. More than 25,000 of the V-1s were produced by the Fieseler Aircraft Company. With this high number, the V-1 is the most produced unmanned fighter aircraft in history, excluding modern hand-launched platforms. The aircraft has flexibility in launching both from the ground and from the air. A powerful pneumatic catapult (catapult) system is used, which is also used in many of today's unmanned aircraft systems. The V-1 was also the world's first jet-powered drone, weighing nearly 5000 pounds with an 1800-pound warhead. Operationally, the V-1 was mainly used in ground-launched rail systems. The V-1 was launched, albeit in small numbers, from the Heinkel 111 bomber, making it the world's first air-launched unmanned aircraft. More than 10,000 V-1s were launched at Allied cities and military targets, killing approximately 7,000 people (Barnhart et al., 2012).

In the period from the first successful UAV flight at the end of the First World War to the end of the Second World War, UAVs had been on the stage of history as weapons-carrying platforms or as target drones. During the Cold War, the purpose of using UAVs changed to reconnaissance, surveillance and deception.

## 2.4 Cold War (1947-1991) and Post-Cold War Period

The Cold War is a period of geopolitical tension between the United States and the Soviet Union and their allies in the post-World War II period. This period lasted until 1991, when the Soviet Union collapsed. The United States and the Soviet Union continued to arm themselves in this period, but because they avoided a hot war, this period was called the cold war. They have supported opposing sides in conflicts in different parts of the world, such as Korea and Vietnam.

Target and reconnaissance unmanned aerial vehicles (UAVs) were both highly researched and developed during the Cold War era, and a variety of UAVs were created by various companies. There are currently a few problems that led to confusion among the UAV types. The US Land Forces or Navy assigned multiple codes to the same UAVs after the manufacturers produced them with a code or name. Due to this, the same UAV had been given three different names. Another problem was that the Forces, who changed the UAV code system over time, came up with a new coding system in place of the preexisting UAV nomenclature, which led to additional confusion.

The development of OQ series UAVs produced by the Radioplane company during the Second World War continued at the beginning of the Cold War era. The OQ-19 Shelduck target drone is equipped with a camera for reconnaissance. The OQ-19 pioneered the current intelligence, surveillance and reconnaissance roles for the UAV (Whitmore, 2016).

In 1948, Ryan Aeronautical Company developed a target drone called the Q-2 "Firebee". The transformation of the drone mission from target drone to reconnaissance drone took nearly 10 years. In 1958, Ryan Aeronautical Company developed the Q-2C Firebee, the first drone developed specifically for reconnaissance missions. The Firebee's J69-T-19 turbojet engine provided an operating ceiling of 60,000 ft. and a range of 800 miles (Blom, 2010). Ryan Aeronautical converted some of the standard Firebee training targets into reconnaissance UAVs and named them the 147A "Firefly". Renamed Lightning Bug, Firefly has been significantly modified to convert from a target UAV to a reconnaissance UAV. In the early 1960s, Ryan Aeronautical designed and developed more than 20 versions of the famous Lightning Bug unmanned subsonic target drone (Keane and Carr, 2013). The Lightning Bug is the grandfather of the Reaper and Predator UAVs in use today. The AQM-34 and BQM-34 series, developed by Ryan Aeronautical Company as a continuation of the Firebee series, were actively used in Vietnam.

During the Cold War, the work of the US Air Force with UAV technology had the most lasting effects. Many types of UAVs had been introduced for different purposes. McDonnel Douglas designed the GAM-72 Quail as a decoy drone in 1961. Jet-powered Quail, launched from the B-52, was intended to confuse the Soviet integrated air defense system (Zaloga, 2008). Quail's purpose was to render the integrated air defense system inoperable in order to increase the survivability of the bomber force as Soviet Surface-to-Air Missile (SAM) systems advanced. But Soviet radar technology has gotten to the point where it can easily identify real bombers with decoy drones. As of this moment, 492 Quails in the inventory of the US Air Force ceased to work (Sloggett, 2015).

## 3. COMPONENTS OF UAS

Unmanned aerial vehicles (UAVs) can't fly or carry out missions by themselves. To fly or become operational, a UAV needs a system that combines many aspects or components. There should be an interdisciplinary compatibility and system management approach because the system's components are composed of individuals from many disciplines.

The components in the UAV system are as follows:

- Unmanned Aerial Vehicle
- Command and Control Element (Ground Control Station)
- Communication Data Link (Ground Data Terminal)
- Payload (Camera, Arms, Weapons, etc. according to the mission requirements)

- Takeoff and Landing Element
- Human Element
- Image Transfer Station
- Ground Support Equipment

The UAS is frequently stated without taking into account the fact that it is a system that functions harmoniously and complementarily. UAV is the word used to describe the flying machine. Unlike manned aircraft, which can fly with just a pilot at the controls, UAV systems do not have all the components on the aircraft. Four components must be present for an aircraft to fly in a UAV system: UAV, UAV Pilot, Ground Data Terminal Element, Command Control Element. The absence of any of these elements will hinder the flight of the UAV. For this reason, it is appropriate to express it as a system. The UAV is the system's primary component.

Instead of the term "UAV," the term "Remotely Piloted Aircraft/ Vehicle" has become more used in recent years. The guidelines and documentation that are available in the UK use the term remotely piloted aircraft system (RPAS). It is estimated that the main reason for this is the thought that the expression unmanned will have different connotations in UAS operations, where human-intensive activities are carried out.

## 4. CLASSIFICATION OF UAS

In the classification of UAVs, various classifications have been made by different countries, organizations and military forces with different approaches. There isn't presently a set standard for classifying UAVs. UAVs are categorized according to a number of characteristics, including their role, size, range, weight, endurance, maximum altitude, wing load, engine type, and command and control or data transfer methods. As can be observed, there are many different categories that UAVs have undergone, mostly based on their unique numerical properties or application areas or roles. It will also become clear that several of these categories cross over with one another.

## 4.1 UAVs by Uses and Roles

UAV systems are divided into two as military and civilian, according to their general use or role.

According to their intended function rather than their characteristics, UAVs classified for use in the military can often be separated into four classes:

- Target UAVs are primarily used to train air defense personnel,
- Non-armed and non-munitions UAVs are made to gather intelligence, surveillance, and reconnaissance (ISR) data,
- Lethal Armed UAVs or Unmanned Combatant/Assault Air Vehicles (UCAV) are made to use ammunition, bombs, rockets, or other weapons while collecting ISR data,
- Striking UAVs, also known as kamikaze UAVs, produced as an intelligent ammunition system.

It is well known that UAVs with military applications are also utilized in numerous other contexts (such as for forward surveillance, target detection, or relay communication), which are made possible by the payloads on the UAVs. UAVs intended for military usage are primarily created for the aforementioned four uses.

UAVs have a wide range of applications that depend on the payloads mounted on them, from recreational use to agricultural spraying to monitoring forest fires to the logistic support they offer in disasters.

## 4.2 UAVs by Platform

Even though UAV platforms are typically divided into four categories: fixed-wing, rotary-wing, flapping UAVs, and balloons (Gupta et al., 2013), it is vital to think of both fixed and rotary-wing UAVs as a separate platform. One of the following five categories best describes UAV platforms:

a. ***Fixed-Wing UAVs:*** In general, they are UAVs that need a runway for take-off and landing. Some of the large UAV systems are designed to take off from a catapult. Depending on the landing and take-off configuration, some of these UAVs can land on the runway as well as land with a parachute. Fixed-wing mini UAVs are generally launched from the shoulder, land on the fuselage or complete the landing with a parachute. Fixed-wing UAVs, other than mini-UAVs, have higher cruising speed and endurance compared to rotary-wing platforms.

b. ***Rotary-Wing UAVs:*** These UAVs that perform vertical take-off and landing (VTOL-Vertical Take Off and Landing) are also called rotorcraft. These UAVs, which do not need any runway due to their vertical take-off and landing capability, also have advantages such as hovering and high maneuverability. These capabilities are useful for many tasks in both military and civilian applications. A rotorcraft UAV can have different configurations, with main and tail rotors (found in a conventional helicopter), co-rotors, tandem rotors, and multi-rotors.

c. ***Both Fixed and Rotary Wing UAVs:*** These UAVs behave in a fixed-wing manner while cruising and are capable of vertical takeoff and landing thanks to their rotary wings. They combine the benefits of helicopters with fixed-wing aircraft. They can fly farther than helicopters because of the lifting force produced by the fixed wing while cruising, yet not requiring a runway for takeoff and landing due to their rotating wings. The term Vertical Takeoff and Landing Fixed Wing UAV is also used for these UAV platforms. Many of these designs can take off and land vertically in addition to using runways.

d. ***Flapping-Wing UAVs:*** Developed with adaptable, small wings, they are inspired by birds and flying insects. Additionally, there are other hybrid and convertible designs (such as the RoboRaven and the DeFly) that can fly like an airplane, tilt their rotors and fuselage, and take off and land vertically.

e. ***Balloons (Blimps):*** Balloons, which are lighter than air and have a long endurance in the air, are usually large in size and fly at low speed.

## 4.3 UAVs by Size

UAVs of various sizes have been created in recent years as a result of the shrinking of electronic components such sensors, microprocessors, batteries, and navigation systems. UAVs come in a variety of sizes, from

*Table 1. UAVs by size*

| UAV Class | Wing Span | Weight |
|---|---|---|
| **Large UAV** | 8-61 m | 800-15000 kg |
| **Medium UAV** | 2-8 m | 5-800 kg |
| **Small/Mini UAV** | 1-2 m | 2-5 kg |
| **Micro Air Vehicle (MAV)** | 15 cm-1 m | 50 g- 2 kg |
| **Nano Air Vehicle (NAV)** | 2.5-15 cm | 3-50 g |
| **Pico Air Vehicle (PAV)** | 0.25-2.5 cm | 0.5-3 g |
| **Smart Dust (SD)** | 1 mm-0.25 cm | 0.005-0.5 g |

the large fixed-wing unmanned aerial vehicle to the diminutive Smart Dust (SD), which is made up of numerous tiny micro-electro-mechanical systems, such as sensors or robots (Hassanalian & Abdelkefi, 2017).

Based on and slightly modifying the spectrum created by Hassanalian and Abdelkefi (2017), UAVs can be classified according to their size as in Table 1.

What is noteworthy here is that as the size decreases, only the term Aerial Vehicle is used instead of Unmanned Aerial Vehicle, as it is unnecessary to use the term unmanned.

## 4.4 UAVs by Take-Off Weight

Take-off weights are typically used as the primary criterion when classifying UAVs by countries, international institutions, and various organizations. They also employed UAV characteristics like altitude, size, and speed as supporting variables in this classification.

Many UAV weight classifications are erroneously designated by including size, altitude, or operational use (Weibel & Hansman, 2004) criteria. Whether a UAV is micro, mini or small is a criterion related to its size, not its weight. In other words, small, mini or micro UAVs are lightweight UAVs. Then, if weight is the criterion in classification, it should be classified as light or heavy UAV. Arjomandi et al. (2006)'s weight scale was used to establish the categorization in Table 2, and the names of the weight classifications were provided.

*Table 2. UAVs by take-off weight*

| UAV Class | Weight | Example |
|---|---|---|
| Very Heavy | >2000 kg | Global Hawk |
| Heavy | 200 kg > weight ≥2000 kg | Hunter, Predator, Heron, Searcher, A-160, ANKA, Bayraktar TB2 |
| Medium | 50 kg > weight ≥ 200 kg | Raven, Shadow, Phoenix |
| Light | 5 kg > weight ≥ 50 kg | Luna, Finder, RPO Midget |
| Very Light | 50 gr < weight ≤ 5 kg | Dragon Eye, Pointer |
| Ultra-Light | ≤ 50 gr | Nano, Pico Air Vehicles and Smart Dust |

## 4.5 UAVs by Altitude and Endurance

There are three levels of altitude at which UAVs can fly: low, medium, and high. It is conceivable to break down this classification into its own subcategories, although UAVs that fly at low and medium altitudes won't be given this distinction. It is helpful to break down the high altitude into subcategories because the tip will continue to be open in the atmosphere.

UAV classification by altitude is as follows (Korchenko & Illyash, 2013):

- Low Altitude: 0-5,000 ft.
- Medium Altitude: 5,000-40,000 ft.
- High Altitude: >40,000 ft.
    - Stratosphere: 40,000-50,000 ft.
    - Suborbital: 50,000-65,000 ft.
    - Special high altitude: 65,000-82,000 ft.
    - Orbit: >82,000 ft.

The endurance (length of time spent in the air) can be divided into short, medium, and long terms. The endurance of UAVs is as follows (Arjomandi et al., 2006):

- Short-term endurance: < 5 hours
- Medium-term endurance: 5-24 hours
- Long-term endurance: > 24 hours

UAVs fall into categories like MALE (Medium Altitude Long Endurance) or HALE (High Altitude Long Endurance), according to the literature. Here, it's important to remember that a classification was created by combining two factors: altitude and endurance (time spent in the air). These categories are as follows (Watts et al., 2012):

- **Low Altitude Short Endurance (LASE):** The most important factor for the practical and mobile use of UAVs is whether they require take-off and landing runways.
  - **No Takeoff/Landing Runway Required:** Most of the UAV systems with Vertical Take-Off and Landing (VTOL) capability are under this classification due to their short stay in the air and flying at low altitude. e-VTOL UAV systems to be used in Urban Air Mobility and fixed-wing mini and small UAVs launched from the catapult and manually launched are also included in this group.
  - **Takeoff/Landing Runway Required:** These fixed-wing UAVs have become slightly heavier and wider than small UAVs as a result of increasing their capabilities. They are capable of staying in the air for several hours and require a runway for takeoff/landing. These systems can remain airborne for several hours at an altitude of approximately 5000 ft.
- **Low Altitude Long Endurance (LALE):** These UAVs are capable of staying in the air for a long time at an altitude of up to 5000 ft., carrying a few kg of payload.
- **Medium Altitude Long Endurance (MALE):** These UAVs are much larger than the low-altitude UAV classes and operate at altitudes of up to 40,000 ft in flights that last for hours and are hundreds of kilometers away.
- **High Altitude Long Endurance (HALE):** The largest and most intricate UAV system exists today. They are larger than manned general aviation aircraft. Under satellite control, these UAVs may fly thousands of kilometers while flying at altitudes of 65,000 feet or higher. These UAVs have over 30 hours of flight records.

## 4.6 UAVs by Range

The ranges/flight radii that UAVs can fly are of great importance in mission planning. UAVs can be classified under five classes according to their range (Korchenko & Illyash, 2013):

Close Range: < 40 km.
Short Range: < 70 km.
Medium Range: <300 km.
Long Range: < 1500 km.
Long Range / Long Endurance: > 1500 km.

## 4.7 UAVs by Control Types

The basic principle of the UAV system is to remove the operator from the cockpit. Therefore, control of the aircraft must be done by other means (Gupta et al., 2013). With advancing technology every day, control in UAV systems is transforming more and more into an autonomous structure. There is a range from the situation where the pilot can fully intervene from the ground to fully autonomous flight. There are three main forms of control an operator can exercise over the aircraft:

- Remote/Ground control,
- Semi-autonomous,
- Autonomous.

Remotely controlled UAVs require constant input from the operator or pilot. These remote-controlled UAVs are essentially radio-controlled aircraft that employ fundamental concepts similar to remote-controlled UAVs used for recreational purposes. Fully remote-controlled UAVs are quite uncommon in the present world. Artificial intelligence and autonomous systems are increasingly being used in unmanned aviation industry.

In semi-autonomous systems, full control of the pilot is required at certain stages of the flight. The pilot must assume full control of the aircraft before the flight, during takeoff and landing, and when operating near the aerodrome. While the UAV is in the air, an autopilot function can be engaged and follow a set of pre-programmed waypoints. The mechanical landing and take-off mechanism at the control station allows some semi-autonomous UAV systems (such as the GNAT-750 and I-GNAT) to be operated by a single pilot, but other semi-autonomous UAV systems (such as Aerostar) may need an internal and external pilot.

At the other end of the UAVs control spectrum are fully autonomous systems. It is based on the principle that the UAV performs the flight

*Table 3. UAV categories*

| Category | Subcategory or UAV Systems |
|---|---|
| Tactical | Smart Dust, Pico, Nano, Micro, Mini, Close Range |
| Operative-Tactical | Short Range |
| Operative | Medium Range, Low Altitude Deep Penetration |
| Operative-Strategic | LALE, MALE |
| Strategic | HALE |
| Special Use | Unmanned Combat Aerial Vehicle (UCAV), Lethal, Decoy, Stratosphere, Exostratosphere, Space |

as pre-programmed, without any human input from take-off to landing. An autonomous UAV can monitor and evaluate its status, whether the flight is sound or not, and the command and control elements on the aircraft within programmed limits. An advanced and superior autopilot performs almost the entire mission on a programmed flight path without any human intervention. The pilot's task during the flight is to watch and follow. In autonomous systems, the entire flight is performed by the autopilot on the aircraft.

## 4.8 UAV Categories

In light of all the aforementioned classifications, UAV systems are grouped under six categories. These are tactical, operative-tactical, operative, operative-strategic, strategic and special purpose use. Table 3 shows which UAV systems fall under these categories (Korchenko & Illyash, 2013):

## 4.9 NATO Classification

NATO's classification of UAVs is absolutely a military classification, given that NATO is a military organization. Although there is no universally acknowledged classification for UAVs, NATO's classification is commonly accepted and used with minor changes by most of the countries.

Although Table 4 includes the category of UAVs, their assigned level, operational altitude, mission radius, and the command level they support,

*Table 4. NATO UAS classification*

| Class | Category | Employment Level | Operation Altitude | Mission Radius | Primary Supported Commander | Example UAS |
|---|---|---|---|---|---|---|
| III (>600kg) | Strike/ Combat | Strategic/ National | Up to 65000 ft. (Mean Sea Level-MSL) | Unlimited | Theatre | Reaper |
| | HALE | Strategic/ National | Up to 65000 ft. (MSL) | Unlimited | Theatre | Global Hawk |
| | MALE | Operational/ Theatre | Up to 45000 ft. (MSL) | Unlimited | Joint Task Force | Heron |
| II (150kg-600kg) | Tactical | Tactical Formation | Up to 10000 ft. (Above Ground Level-AGL) | 200 km | Brigade | Hermes 450 |
| I (<150kg) | Small (>20kg) | Tactical Unit | Up to 5000 ft. (AGL) | 50 km | Regiment, Battalion | Scan Eagle |
| | Mini (2-20 kg) | Tactical Subunit | Up to 3000 ft. (AGL) | Up to 25 km | Company, Platoon, Squad | Skylark |
| | Micro (<2kg) | Tactical Subunit | Up to 200 ft. (AGL) | Up to 5 km | Platoon, Squad | Black Widow |

Source: NATO UAS Classification Guide. September 2009 Joint Capability Group on Unmanned Aerial Vehicles (JCGUAV) meeting

the weight of the UAVs serves as the primary criterion for classification. One of the major problems is that UAVs weighing less than 150 kg are not subject to NATO certification regulations.

## 5. CLASH OF STRATEGIES, CONCEPTS, AND TECHNOLOGY ON UAS AND ARISING CHALLENGES

Concept development and strategy creation efforts by aviation authorities around the world lag behind the rapid and superior developments in Unmanned Aerial System (UAS) technologies in recent years.

In 2005, The Secretary General was asked by the Air Navigation Commission to consult with a select group of States and international organizations regarding the following topics: current and anticipated international civil unmanned aerial vehicle (UAV) activities in civil airspace; procedures to eliminate the risk that UAVs operated as State aircraft pose to civil aircraft; and potential processes for the issuance of special operating authorizations for international civil UAV operations. Then, two informal meeting were held on UAVs in Montreal in 2006 and in Palm Coast in 2007. At the first informal exploratory meeting on UAVs held in Montreal, a very interesting agreement conclusion emerged. Its goal was to ascertain the potential contribution of ICAO to the establishment of UAV regulations. The group came to the conclusion that while there would eventually be a wide range of technical and performance requirements and specifications, only certain of them would require becoming ICAO SARPs. Additionally, it was found that ICAO was not the best organization to head the endeavor to create such specifications. However, it was decided that ICAO should serve as a focal point and that there was a need for harmonizing terms, approaches, and concepts with regard to the regulatory framework. In the next informal meeting in Palm Coast in 2007, the task of working on technical regulations and standards for UAV operations was assigned to a joint committee composed of two organizations, Radio Technical Commission for Aeronautics (RTCA) and European Organisation for Civil Aviation Equipment (EUROCAE) (ICAO, 2011). Although issues with the integration of UAVs into the airspace started to emerge, the UAV was not considered to be an actual aircraft, and the issues were attempted to be dealt with different organizations rather than the ICAO. Unfortunately, despite improvements to various models and many simulations being run, even ICAO, NASA, and EASA today are unable to provide an absolute answer on how to integrate UAVs into the airspace.

With the application of artificial intelligence and smart technologies to UAS and widespread use of UAS in miscellaneous fields, aviation authorities have abruptly faced great challenges in the integration of UAVs into the airspace, air traffic regulations and concept development. Smart technologies and artificial intelligence transformed the UAS into different forms like flying swarm in the airspace and Urban Air Mobility (UAM) which comprises passenger and cargo transportation without pilot in metropolitan cities.

According to EASA (2023): "*UAM is a new safe, secure and more sustainable air transportation system for passengers and cargo in urban environments, enabled by new technologies and integrated into multimodal transportation systems. The transportation is performed by electric aircraft taking off and landing vertically, remotely piloted or with a pilot on board.*" By 2025, UAM commercial operations are expected to start in EU cities. Although EASA (2023) claims 340 million people who live in EU cities are expected to experience UAM by 2030, this projection doesn't seem realistic as many problems are still unsolved. EASA (2023) counted the benefits of UAM as faster, safer, and cleaner mobility, positive economic impact with 90.000 job opportunity, and gaining the 31% of global market as a leader.

Federal Aviation Administration (FAA) NextGen Office released the UAM Concept of Operations version 2.0 in 2023 which took over the initial publication of 2020 document after having feedback from Federal Aviation Administration (FAA), National Aeronautics and Space Administration (NASA), and industry partners. The aim of this document is to mature the concept of Advanced Air Mobility (AAM) which provides transportation between urban, local, regional, and intraregional areas where are poorly served and/or not previously served by using aircraft equipped with the latest and most innovative technologies. UAM is a subset of AAM. While UAM flight operations are in and around urban areas, AAM provides a wide range of passenger, cargo and other operations within and between urban and rural environments. Since traditional Air Traffic Services (ATS) is insufficient in these operations, it is supported by Extensible Traffic Management (xTM) which is still partially under discussion (FAA, 2023). Beside these efforts of FAA and NASA for conceptualization, The European Union Aviation Safety Agency (EASA) have also had some regulatory efforts on UAM and prepared some reports on societal acceptance of UAM.

Despite all these developments, there is no determined UAV operating model or a strategy related to UAV operation in the literature and in reality. Probably as a result of existing operational experience, different business models will be created for highly differentiated UAV types. The most notable reason for this situation is that UAV manufacturers have created a monopoly in the sector from training to operation. An insight has prevailed that no business can be founded without manufacturers. This is also one of the great challenges faced by UAS and UAM businesses. Any kind of UAS business will depend on the civil application areas while military usage of UAS is determined according to the need of the operation. The structures of the UAS businesses are ambiguous yet.

After all, technological inputs and innovations bring their own risks and thus different challenges arise. Various prototype UAVs for UAM to carry passengers and freight have been built by numerous firms. Pilot cities like Guangzhou and Dubai were chosen to experience UAM flights, and the implementation process was tracked. The most conspicuous challenges for UAV and UAM might be listed as follows and this list might be extended:

- Aircraft airworthiness,
- Integration to airspace,
- Air traffic control,
- Pilot and technician licensing,
- Charge stations,
- Societal acceptance of UAM, etc.

## 6. CONCLUSION

To the extent that human beings force their imagination, they take smart technologies and artificial intelligence to a higher level every day. UAS has undergone tremendous technological evolution since its inception. US officials have stated at every opportunity that the F-35 is the last manned fighter aircraft produced. This shows that, thanks to robotics, artificial intelligence, and learning machines, many people will be unemployed in the aviation industry. UAVs are now widely employed in civilian life and are no longer restricted to use on battlefields. Pilots will no longer be pilots of manned aircraft, but of unmanned aircraft.

Unfortunately, strategies, concepts and regulations have lagged behind developments in UAV technology. Therefore, great challenges arise in transitioning to the field of application and prototypes do not appear in the airspace by starting mass production. In other words, the airspace is not yet ready for traffic that may occur over cities, as in road traffic. Aviation authorities have been delayed and unsuccessful in bringing the produced UAVs into the airspace for AAM and UAM.

## REFERENCES

Arjomandi, M., Agostino, S., Mammone, M., Nelson, M., & Zhou, T. (2006). Classification of unmanned aerial vehicles. Report for Mechanical Engineering class, University of Adelaide, Adelaide, Australia, 1-48.

Barnhart, R. K., Hottman, S. B., Marshall, D. M., & Shappee, E. (2012). Introduction to Unmanned Aircraft Systems. CRC Press, Taylor & Francis Group, FL, USA.

Blom, J. D. (2010). *Unmanned Aerial Systems: A Historical Perspective*. Combat Studies Institute Press.

Cir, I. C. A. O. 328 (2011). Unmanned Aircraft Systems. Springer.

Coletta, P. E. (1987). *Patrick N. L. Bellinger and U.S. Naval Aviation*. University Press of America, Inc.

EASA. (2023). *Urban Air Mobility (UAM)*. EASA. https://www.easa.europa.eu/en/light/topics/urban-air-mobility-uam

FAA (2023). *Urban Air Mobility (UAM) Version 2.0-Concept of Operations*. FAA.

Fahlstrom, P. G., & Gleason, T. J. (2012). *Introduction to UAV Systems* (4th ed.). John Wiley & Sons, Ltd.

Gupta, S. J., Ghonge, M. M., & Jawandhiya, P. M. (2013). Review of Unmanned Aircraft System (UAS). [IJARCET]. *International Journal of Advanced Research in Computer Engineering and Technology*, *2*(4), 1646–1658.

Hassanalian, M., & Abdelkef, A. (2017). Classifications, applications, and design challenges of drones: A review. *Progress in Aerospace Sciences*, *91*, 99–131. doi:10.1016/j.paerosci.2017.04.003

Keane, J. F., & Carr, S. S. (2013). A Brief History of Early Unmanned Aircraft. *Johns Hopkins APL Technical Digest*, *32*(3), 558–571.

Korchenko, A. G. & ve Illyash, O.S. (2013). The Generalized Classification of Unmanned Air Vehicles. *IEEE 2nd International Conference Actual Problems of Unmanned Air Vehicles Developments*. IEEE.

Monash University Hargrave-Andrew Library. (2006). *Remote Piloted Aerial Vehicles: An Anthology*. Monash University. https://www.ctie.monash.edu/hargrave/rpav_home.html#Beginnings

NATO UAS Classification Guide. September 2009 *Joint Capability Group on Unmanned Aerial Vehicles (JCGUAV) meeting*. NATO.

Newcome, L. R. (2004). *Unmanned Aviation: A Brief History of Unmanned Aerial Vehicles*. American Institute of Aeronautics and Astronautics. doi:10.2514/4.868894

Singer, P. W. (2009). *Wired for War: The Robotics Revolution and Conflict in the Twenty-First Century*. Penguin Press.

Sloggett, D. (2015). *Drone Warfare: The Development of Unmanned Aerial Conflict*. Skyhorse Publishing.

Watts, A. C., Ambrosia, V. G., & Hinkley, E. A. (2012). Unmanned Aircraft Systems in Remote Sensing and Scientific Research: Classification and Considerations of Use. *Remote Sensing (Basel)*, *4*(6), 1671–1692. doi:10.3390/rs4061671

Weibel, R. E., & Hansman, R. J. (2004). Safety considerations for operation of different classes of UAVs in the NAS. *Proceedings of the 4th Aviation Technology, Integration and Operations Forum, AIAA 3rd Unmanned Unlimited Technical Conference, Workshop and Exhibit*. IEEE.

Whitmore, B. A. (2016). *Evolution of Unmanned Aerial Warfare: A Historical Look at Remote Airpower—A Case Study in Innovation*. [Thesis, U.S. Army Command and General Staff College, Fort Leavenworth, KS].

Zaloga, S. (2008). *Unmanned Aerial Vehicles: Robotic Air Warfare, 1917-2007*. Osprey Publishing.

# Chapter 7
# The Synergy Between Tourism, Aviation, and Digital Transformation:
## A Path to Seamless Travel Experiences

**Gonca Güzel Şahin**
*Atilim University, Turkey*

## ABSTRACT

*This chapter explores the profound impact of digital transformation on the tourism and aviation sectors, highlighting the necessity for a comprehensive digital strategy in response to global changes. The tourism industry, in particular, is undergoing a paradigm shift, leveraging technological advancements such as mobile applications, social media, and digital platforms to enhance customer satisfaction, experience, and overall competitiveness. Within the aviation sector, digital developments span smart maintenance, risk management, air traffic optimization, and customer relations, all aimed at improving operational efficiency and passenger experience. The interconnectedness of tourism and aviation is evident in the increasing role of air travel, accounting for over 52% of global tourist movements. The chapter delves into the historical evolution of this symbiotic relationship, emphasizing the pivotal role of innovations, increased connectivity, and affordability in fostering tourism growth.*

DOI: 10.4018/979-8-3693-0732-8.ch007

## 1. INTRODUCTION

The tourism sector has increasingly emphasized the use of technology in response to global changes over time. In conjunction with economic development, a digital transformation plan that also addresses environmental factors has become a necessity for the tourism sector. With the process of digital transformation, the effort to maintain customer satisfaction, experience, and a competitive environment has now become indispensable for businesses. Industry 4.0 has introduced a new lifestyle in all aspects of life, leading us to use technological devices such as the Internet, tablets, robots, and other technological tools. The primary goal of using technology in tourism is to make people comfortable in planning their travels and ensure the reliability of their journeys. New generation technological advancements aim to facilitate customers' travels as well as enhance the efficiency of businesses in terms of money, time, and effort across various sectors. The introduction of mobile application programs for communication with customers, navigation applications, and mobile payment systems has improved the quality of service in the tourism sector, making travel activities more accessible and increasing customer satisfaction. In addition to digital facilitators in customer experience, the mobile compatibility of social media platforms and other sharing platforms is also crucial in shaping the customer experience.

Airlines within the tourism sector also heavily benefit from digital developments while performing their core functions and managing supply, maintenance, operations, and customer relations. In the aviation sector, digital transformation encompasses various applications, including smart maintenance, risk management, air traffic optimization, customer satisfaction, cost reduction, revenue management, performance metrics, cost control and verification, cargo load control aimed at increasing fleet capacity, and ticket price adjustments, providing passengers with a tailored travel experience, and managing airport performance. The aviation sector aims to increase profitability through digital applications while also making it almost essential to maintain its presence in the sector in terms of customer satisfaction and competition. Considering both internal and external dynamics, the aviation sector significantly differs from other sectors in terms of innovation and renewal, making substantial use of digital applications in the process. In this context, this study focuses on the digital transformation and digital application areas in the aviation sector.

Innovations in technology, improvements in working conditions, increased vacation days, the establishment of travel agencies, and the growth of tourism investments have all encouraged people to travel, leading to significant advancements in tourism since the 1960s. The continuous development of the relationship between tourism and airlines from its inception to the present day can be explained by both sectors being open to innovations and developments and being interconnected. The increased connectivity between destinations and the relative ease and affordability of transportation options have played a pivotal role in the global development of tourism. More than 52% of global tourist movements are carried out by air travel. Especially for destinations that are geographically distant from each other, air transportation plays a fundamental and accelerative role in the realization of tourism.

In the years following the increase in demand in tourism, unprecedented developments have taken place in international air transportation. Advances in airline technology, low fuel prices, the use of jet and wide-body aircraft, and international air traffic have had a positive impact on tourism. Air travel, which has shown parallel development to mass tourism, is a type of transportation that particularly enhances long-distance travel. Organized trips, including ground services and often referred to as package tours, have become the most preferred way of traveling for people. Therefore, there is a mutual interdependence between tourism and aviation.

Current trends in tourism are making aviation an increasingly growing sector for various reasons. Among these reasons, the increase in the number of people traveling by plane, the increase in individual travel, and the increase in the average travel distance per flight can be cited as examples. Regulations related to the operation of the sector have led to a significant decrease in ticket prices in air transportation. Low ticket prices have increased the demand for air travel among consumers who want to save time and travel more comfortably, leading to more frequent flights and significant growth in air transportation. This situation has resulted in increased mobility, accessibility, and improvements in service quality from both the perspective of the aviation sector and users. Aviation plays a central role in ensuring sustainability in tourism. Therefore, in terms of sustainable tourism, it can be said that sustainable transportation systems have a significant impact. The current and future growth of the aviation sector also implies the increasing challenges in managing ecological balance, both at the national and international levels. These challenges include issues such as the impact of emissions produced by aircraft on global warming and the management of waste generated during aviation operations.

This chapter explores how digital technologies are revolutionizing both the tourism and aviation industries and creating synergies that can lead to seamless and enhanced travel experiences for passengers. It examines how the integration of digital innovation is reshaping the way people plan, book, and experience their journeys.

## 1.1 Digital Transformation in Tourism

The impact of digital technologies on the tourism industry has been profound and transformative. From the invention of the wheel to the development of jet engines, the discovery of language and writing, and the advancements in information and communication technologies, various technological breakthroughs have influenced and reshaped the tourism sector. In this context, tourist products have not only been the elements that make up or shape tourism but have also emerged as a result of the evolving tourism landscape (Kozak et al., 2013: 10). Evolving technology and the use of social platforms have empowered consumers, putting them in a stronger position. With digitalization, changes have occurred in consumer wants and demands. The tourism sector has had to adapt to this transformation. Digitalization in tourism is a result of the use of modern information technologies within the industry. It has revolutionized the way travelers plan and experience their trips and has compelled businesses in the tourism sector to adapt to these changing consumer behaviors and expectations. The advancement of technology and the use of social platforms have indeed empowered consumers, placing them in a stronger position. With the digitization of various aspects of life, changes in consumer preferences and demands have become evident. The tourism sector has had to adapt to these changes. Digitalization in the tourism industry has emerged as a result of the utilization of modern information Technologies (Happ & Horvath, 2018: 10). The increasing environmental awareness and the growing importance of sustainability have led to the emergence of a tourism phenomenon that extends from the real world to virtual and immersive experiences. In fact, technology is being extensively leveraged in this form of tourism (Urry, 1999: 206). The phenomenon of digitization, which will continue to advance in the future, could potentially lead to individuals being able to take vacations or travel without leaving their homes, possibly reaching the realms of augmented reality or virtual reality experiences. Digital tourism is defined as the use of digital technologies to enrich the tourist experience (Adeola & Evans, 2019: 190). Throughout history, the tourism industry has demonstrated a remarkable ability to adapt to changing circumstances and technological advancements. It has evolved to meet the evolving needs and desires of travelers. In the face of the ongoing digital transformation, it is expected that businesses within the tourism industry will also demonstrate a high level of adaptability and embrace digital tourism (Urry 2009: 219). In their study, they stated that the concept of smart tourism was introduced by Gordon Phillips in a conference held in the year 2000. Planning tourism businesses, developing tourism products, and thus marketing them are all part of smart tourism with a holistic approach. In this context, smart tourism is expressed as a sustainable and future-oriented concept. Enhancing tourists' tourism experiences and better managing the desires of destination stakeholders require the use of technological infrastructure (Li et al., 2017: 294). The definition of smart tourism made by Geoffrey Lipman, the Deputy Secretary-General of UNWTO, expresses it as "clean, green, ethical, and quality" at every level of the service chain. Improving management and governance, facilitating the accessibility of products and services, enriching the tourist experience, and ultimately enhancing the competitiveness of tourism businesses and destinations with other destinations are among the goals of smart tourism through the development of IT infrastructure (Gretzel et al., 2015: 175).

Digitization is an important step for tourism businesses, countries, or regions to enhance their success in the sector (Güven et al., 2019: 2). In recent years, technological advancements and the growth of the internet, along with the proliferation of smartphones and other digital applications, have given rise to the concept of digital travel in advanced countries, transforming it into the tourism industry (Law et al., 2018). With digital tourism, tourism has evolved from being merely local to gaining a global dimension. Comments made by someone on an article or post in any social medium, photos shared on a photo-sharing site, and the entire content of an internet blog all contribute to the creation of social media, with access to all of these occurring in a global online environment. The digitization of tourism demonstrates how important these elements are (Başaran & Mercan, 2019: 2). With digital advancement, the goal has also been to improve the quality of tourism technologies. Using multimedia guides instead of carrying travel plans and maps has a significant impact on tourists' experiences and has become easier for travel reservations (Benyon et al., 2013). In this respect, digital tourism is about supporting the tourism experience with digital tools. The innovation in digital tourism is, in fact, the digital development of tourism, aiming to elevate quality and enhance the tourist experience through the use of technology. The concept of digitization is widely used in the marketing mix through accommodation facilities. There are many studies related to the digitization of tourism. Digital tourism encompasses geographic information systems, remote sensing, global satellite navigation systems, and information technologies. This completeness is a computer system that covers all elements of a digital tourism activity, encompassing the entire process. Real-time digital tourism updates tourism information and services, offering the latest tourism trends to governments, entrepreneurs, and tourists in their most current form (Wu, 2020).

In the literature on smart tourism destinations, there is a significant effort in the industry to use information and communication technologies to improve tourism service processes and increase tourist production (Wang & Li, 2013). The intensive use of information in the tourism sector, along with the increasing interest in information and communication technologies, has made the use of digitization in tourism inevitable (Koo et al., 2015:100). Along with digitization, new conditions have been added to the value chain of tourism activities, the balances of existing conditions have changed, and the number of tasks and operations has decreased (Dredge et al., 2018). The widespread use of mobile devices, especially smartphones and numerous applications, today provides a unique connection and access to the internet from anywhere (Wang & Xiang, 2012: 309).

For many years, the tourism industry has heavily utilized computer programs in ticket reservation and sales processes. As a result of the development of information systems, the internet and websites have become new marketing tools. Tourism, being a global industry, has become a part of the information industry. Consumers wanting to obtain basic information about a product before seeing or using it have facilitated mutual information exchange with tourism facilities. This information acquisition becomes important due to the independence of tourism businesses and consumers from location. Purchasing tourism products without testing or trying them creates a network of information exchange between accommodation facilities and consumers, requiring secure, fast, and reliable information exchange. Using big data technologies, it is possible to determine and manage customers' needs in the tourism sector. Therefore, the role of big data in planning and organizing the tourism sector is important (Esen & Türkay, 2017: 92).

Additionally, customers can obtain the information they need for the selection of tourism-related products and destinations, as well as search for, find, and share this information (Papel & Prideaux, 2016). Due to the popularity of internet usage, many accommodation businesses, transportation companies, and travel agencies have started incorporating internet technologies into their marketing and communication channels (Buhalis & Law, 2008: 611). Advancing technology has brought even more to these digital services. With the global reach of the internet, business managers can instantly access up-to-date information from any geographic location in seconds and enhance innovation in their businesses with frequently discussed artificial intelligence applications, leading to the emergence of the concept of digital tourism (Şanlı et al., 2019: 4). Big data is a crucial component for the concept of digital tourism. Big data is a new concept that allows the identification of different data in different volumes that cannot be processed using traditional database techniques and contains various digital content (Gahi et al., 2016: 953). As a result, businesses that want digital tourism to provide a sustainable competitive advantage should ensure the efficient management of social media, big data, and all digital applications (Mariani et al., 2016: 327). Especially travel agencies can use the data of past tours and current reservation data to make accurate predictions of future demand. They can also combine this information with local events, weather conditions, and flights and use them effectively in marketing efforts (Revfine, 2021).

Travel is a rapidly evolving concept that requires rapid data analysis and quick decision-making in the process. To provide tourists with the services they desire, continuous interaction with them is necessary, highlighting the importance of big data. With the increasing amount of big data, it is believed that new data and methodologies will be created instead of traditional tourism data. For example, call centers that create interfaces with online consumer reviews, loyalty programs connected to reservation history, and accommodation preferences will be combined with social media conversations (Song & Liu, 2017: 19). Tourists, city residents, tourism operators, official tourist boards, destination managers, bloggers, local tourism associations, and all stakeholders create big data through their content, data, videos, photos, messages, text, comments, code, likes, and other interaction activities. Real-time data obtained in this way improves decision-making processes, predicts needs in advance, reveals variability, personalizes and customizes offers, and enhances and enriches the tourist experience. In the past, people had to go to agencies to make holiday reservations, but today, all kinds of services can be done online. People's travel preferences are composed of a variety of personal choices. Where one goes on vacation, where one stays, and even which airline they fly with provide important information about a person. Having knowledge about customer behavior and preferences in online travel choices is valuable not only for relaxing customer travel choices but also for customer satisfaction and sales success for travel agencies. In the world, the tourism industry is using technology to transform itself. As technology advances, reservations and sales have become automated. Travel agencies and tour operators are suitable tools for integrating different forms of interaction through both front-end transactions and online services in travel services and tour operations with the internet of things technology. IoT devices can provide valuable data almost in real-time or in bulk processes. Given the importance of big data and analytics, which are key tools for extracting valuable information from data, the Internet of Things plays a crucial role in improving tourism services for tour operators and travel agencies (Polat & Esen, 2021: 168). The transformation, which began with smart tourism applications, has led to the emergence of the concept of smart tourism destinations, especially with the development and transformation of smart cities (Boes, 2015: 11). Accordingly, businesses in the tourism sector are striving to adopt new technologies in their efforts to become digital (Leung et al., 2013).

## 1.2 Case Studies of Innovative Tourism Companies Leveraging Digital Tools for Customer Engagement

Images of historically and culturally significant buildings in the city are processed with image recognition software and stored in three dimensions. Through QR codes, tourists can interactively view the works located on predefined routes known as "red" and "blue," choose routes for walking accordingly, and view the works (Fino et al., 2015: 340). Virtual reality, which has become part of the tourism industry with digital transformation, is used to help tourists make decisions about destinations or to see a destination early as a technology-focused tourism application. For example, Hotel Zetta in San Francisco has a room where this application can be carried out within its facilities, providing the opportunity for tourists who prefer to travel over time (Atar, 2019: 106). Border robots used to process biometric information during tourists' entry into the country and to perform passport procedures, Henn na Hotel in Japan, where reception procedures are carried out by robots, robot security guards at airports performing facial scanning, and robot travel agencies used by the Amadeus company to understand tourists' travel expectations and develop recommendations can be shown (Türsab and TBV, 2019: 33). Seul Deoksugung and Gyeongbokgung Palaces, Dublin An Post Museum, Paris Louvre Museum, and London British Museum, some of the world's leading heritage and tourism organizations, are developing mobile augmented reality applications (Lee et al., 2015: 478). At Istanbul Modern Museum, visitors are provided with an interactive experience through Bluetooth, and personalized touring programs are recommended to visitors based on their available times, with multi-language support. In addition, the museum's mobile application provides information about the artworks when approached, and allows for various interactions. The Museum Hotel Antakya offers services with its unique status as the first hotel built on a museum. The building houses historical walls dating back to 300 BC, 30,000 historical artifacts, the world's largest single-piece floor mosaic, Roman baths, a Pegasus depiction from the 2nd century, a health center, a ballroom, and five restaurants. The museum includes videos, touchscreen displays, and 3D projections that showcase ancient Antakya stories, offering a balanced relationship between technology and history (The Museum Hotel Antakya, 2020). Hilton Garden Inn Istanbul Golden Horn adopts sustainability principles with its LEED (Leadership in Energy and Environmental Design) green building certification. In addition, the building uses advanced technologies in environmental efficiency, such as recycled construction materials, water-saving systems, lighting, and cooling management (Hilton, 2020). Cloud 7 Hotel in Istanbul does not have a reception. Check-in and check-out processes at the hotel are carried out through a smartphone application. In addition, keyless entry systems are used in hotel rooms, and orders can be placed and services can be obtained through WhatsApp to the staff (Yalçınkaya et al., 2018: 94). Barcelona provides tourists with smart bus stops that offer tourist information and bus arrival times in their own language, as well as USB charging ports for their mobile devices. The city also supports the more active use of bicycles throughout the city and offers a smartphone application service for tracking their locations. Brisbane uses more than 100 signs to provide information to tourists through a mobile application. Amsterdam uses smart signs to provide services to tourists in their own languages and uses the Arena system, which operates through sensors, to better manage tourist crowds. Seoul invests in providing free phones and Wi-Fi services to tourists. Jeju Island in South Korea introduces its BIT infrastructure to offer innovative experiences to tourists and declares itself a smart tourism destination. Sunmoon Lake in Taiwan offers location-based information systems to make travel more convenient for tourists. All of these infrastructure projects enrich physical infrastructure with data to promote innovation, increase the quality of life, and encourage sustainability as parts of larger, coordinated, and strategic investments (Gretzel et al., 2015: 180). In Barcelona, NFC tags and QR codes are provided in tourist areas to allow tourists to easily obtain information using their phones, and data is collected to improve the tourist experience. A study initiated at Sagrada Familia produces congestion maps of tourists, which is planned to be applied to other tourist areas. With the Dubai Token application, tourists can earn tokens by touring or shopping and spend them at affiliated locations. In addition, various applications are used to enhance the tourist experience. Real-time feedback from this data will provide insights into tourist behavior and enable personalized service. TUI, in collaboration with Utrip, an artificial intelligence company based in Seattle, uses artificial intelligence infrastructure to analyze reviews of different destinations from major sources such as Tripadvisor and Yelp and make personalized travel destination recommendations by combining them with TUI destinations. The Red Roof hotel chain found that an average of 92,000 flights were either canceled or delayed in the United States through an analysis of flight data. Using real-time analysis methods, they sent mobile search ads to passengers whose flights were delayed or canceled, resulting in a 60% increase in reservations. Kayak, an online travel agency, offers information through a mobile application using Amazon Alexa voice commands to find and book hotels and flights for users (Türsab & TBV, 2019: 29). Radisson Blu & Spa Istanbul Tuzla offers services with its smart building features, including ventilation, cooling, lighting systems, and hidden Wi-Fi areas. For example, during meetings, participants can communicate their preferences to the staff using tablet applications. Guests' preferences are monitored 24/7 through IP-based televisions and phones, and these preferences are recorded to offer personalized experiences in future travels and make improvements accordingly. The Express Check-in/Checkout application allows guests to easily complete transactions. In addition, the OneTouch mobile application provides guests with access to all information about the hotel (Radisson Blu, 2020). The Dedeman Group aims to provide services tailored to the travel habits of the new generation with Smart By Dedeman. The hotel, built on the theme of "Smart Living," allows guests to explore rooms with virtual reality before arriving at the hotel, making check-in and check-out processes easy with a smart mobile application, and is equipped with gaming areas, state-of-the-art meeting rooms, and smart technologies to provide guests with a new accommodation experience (SmartByDedaman, 2020).. In Turkey, the Turkey Travel Agencies Association (TÜRSAB) has launched the TÜRSAB Roto Project in partnership with the financial technology brand Iyzico. In this project, package tours prepared by many travel agencies in Turkey are offered to customers with secure payment through Iyzico's AI technology-based infrastructure (TÜRSAB, 2022). The Amadeus Information Group has developed a search engine that uses mixed reality technology. For example, users can turn the world to select a destination. If the traveler likes what they see, they can purchase a flight and go through the entire booking process. When selecting a flight, the user can walk around the plane cabin to find the right seat or make a rental car reservation. Users experienced their first mixed reality experience through this search engine, which uses realistic environmental sounds, 360-degree videos, and holograms (Amadeus, 2017). Many travel agencies around the world have introduced mobile applications. Booking.com, Airbnb, Hopper, Expedia, Agoda, Vrbo, Trivago, Hotels.com, Oyo, Tripadvisor, Jolly Tour, and Ets Tour are among the travel agencies with mobile applications (Turizm Press, 2022). In tourism, digital applications are used to measure and access data related to energy savings, water conservation, waste management, and carbon footprint reduction in hotels. The Marriott chain has a robot named Mario that can inform guests in their own languages about general areas, rooms, restaurants, bars, and more. In Japan, at Henna Hotel, all services except security are provided by robots with facial recognition capabilities (Güler, 2019: 431).

## 1.3 Digital Transformation in Aviation

Biometrics is an automatic person identification system that uses physiological characteristics (face, fingerprints, hand geometry, handwriting, iris, retinal, vein, voice). Existing research shows that there are numerous biometric services in airports, including airport security, biometric travel documents, airport access control, biometrics in security, biometrics in baggage claim, airport immigration systems, and seamless travel (Heracleous & Wirtz, 2006). "Smart airport" is a term referring to the intelligent infrastructure of an airport. It is derived from the concepts of "smart cities," "smart buildings," and "smart factories." The global smart airport infrastructure is divided into endpoint devices, which include communication systems, passenger, cargo, and baggage handling, air traffic control systems, security systems, and others. Endpoint devices are further subdivided into sensors, tags, IP phones, and video conferencing in the future (Chesher, et al., 2013).

With a smartwatch, passengers can receive alerts about gate changes or flight delays, and in cases of lateness, they can scan their boarding passes. Since a smartwatch is worn permanently on your wrist, ground service agents can use this channel to broadcast real-time information to passengers beyond traditional visual display monitors, voice announcements, and other mobile devices (Rawassizadeh et al., 2015). Key companies dominating the global smart airport market include IBM Corporation, Cisco, and Siemens AG. Other companies like Raytheon, Amadeus IT Group, and QinetiQ are identified as emerging players in the market. Companies like Honeywell and SITA are considered strong contenders and promising firms in the smart airport market (Angelakis et al., 2017). Kiosks that serve the check-in applications of multiple airlines offer several benefits. Their use provides advantages at airports due to limited space usage, eliminates capital expenditures for airlines since they purchase these services directly from ground service agents, and reduces operational costs for ground service agents by employing fewer ground personnel for passenger control. Passengers save time because they can use any kiosk instead of searching for airline-specific kiosks. Ground service agents establish self-service check-in and support passengers with their staff to encourage the use of self-service check-in kiosks on the landside of entry areas (Wittmer, 2011).

Digital baggage tags are the digital alternatives to traditional paper-based baggage tags. Baggage receives a permanent baggage tag displaying a digital barcode. If the flight plan changes or if the passenger revisits, airlines or ground service agents can remotely update this barcode. Along with a tracking device inside the baggage, passengers can track their baggage in real-time using a smartphone (Want, 2006). Self-service check-in kiosks are computer terminals that eliminate the need for ground personnel for passenger entry. In the past, airlines used kiosks exclusively for their own passengers, but today, kiosks serve the check-in applications of multiple airlines (Howes, 2006). The Self-Boarding system shortens passenger wait times, enabling passengers to move to the aircraft boarding area more quickly. By only allowing the passage of individuals with a valid boarding pass, it contributes to ensuring the security of the aircraft boarding area and prevents unnecessary crowds by blocking the passage of individuals other than passengers. Additionally, airport ground personnel have more time to attend to passengers' needs (Airport Solutions, 2021). Self-Boarding gates automate the boarding process, allowing passengers to access the aircraft using a boarding pass or biometric identity. With the latest sensor technology, it ensures that only those with a valid boarding pass can access the aircraft boarding area (Selfboardinggate, 2022).

Mobile applications in airlines aim to improve customers' travel experiences. In 2020, 68% of airlines introduced AI-powered chatbots. Chatbots at airports not only provide quick access to essential information but also answer passengers' questions. At airports, chatbots provide passengers with 24/7 information about flight details, airline services, airline transportation, baggage information, parking services, general information, special services, and Covid-19 services, allowing passengers to receive this information in their primary languages. (Datasemantics, 2022). The benefits of RFID baggage tags are widely discussed in the existing literature. RFID tags can reduce baggage mishandling, improve baggage tracking, and decrease the number of bags sent on incorrect routes during air transportation (DeVries, 2008). Tags combined with printer printing or electronic baggage tags allow passengers to use fully automated machines to retrieve their tagged baggage without any interaction with ground personnel or airline staff. Some airports have already installed automated baggage drop machines (Jaffer & Timbrell, 2014).

On the other hand, airlines also want to use digital technologies to meet customer demands. According to IATA's global passenger survey report, the younger generation desires features such as making payments with a digital wallet on their smart devices, having in-flight WiFi, and receiving continuous updates and notifications via messages or emails about pre-travel arrangements and requirements (IATA, 2019). Travelers using sensors at airports can track how far their luggage is from them, saving them time. With the help of sensors, tourist locations can be tracked, and promotional messages can be sent to their phones (Revfine, 2021). Airlines have developed mobile applications to enhance customer loyalty, reduce costs, and increase interactions with customers. These mobile apps allow passengers to check in, search for flights, request additional baggage allowances, and quickly perform various services during flights, such as selecting meals (Bingemer, 2013: 41). Quick boarding gates enable passengers to scan their boarding passes themselves. Once the boarding pass is verified, the gates are released, allowing passengers to proceed to the aircraft. In this process, ground personnel are not involved in passenger boarding and are responsible for inspection duties or special cases. This helps reduce the need for ground personnel and further lowers operating costs for ground service agents. Current literature suggests that self-management positively affects customer satisfaction because it significantly reduces processing times at the gate (Jaffer & Timbrell, 2014).

Since the 1980s, the development of these technologies has completely transformed the structure of global tourism. The internet has revolutionized the dissemination of travel information and the way people plan and consume their travels. With the increasing development of information and communication technologies, tourism has become entirely information-based (Buhalis & Law, 2008). In the airline industry, Global Distribution System (GDS) providers and the airline systems they offer play a significant role in the use of digital technologies in various areas. Some of these providers include Sabre, Amadeus, Galileo, Worldspan, Lufthansa, IBM, LIDOSITA, Open Skies Navitaire, NCR/Teradata, Oracle, Airport Technology International, Air Transportation Association, and Inflight Catering Association. Through global distribution systems, airline companies can access ticketing, reservations, sales, airline flight information more quickly, stay updated on flight regulations, make last-minute cancellations or changes, track low fares, access car rental services, and make reservations at accommodation establishments integrated into the system. Many airline companies receive services from one of the four major GDS providers (Bingemer, 2022). The tourism industry, evolving with three industrial revolutions, can be defined as the era of Tourism 4.0 digital transformation. This era incorporates artificial intelligence, the Internet of Things, online booking systems, payment systems, destination management, customer relationship management, smart hotel systems, smart human resources, smart travel agencies, virtual reality that allows tourists to explore and get information about previously unseen destinations, and mobile applications (Şanlıöz et al., 2013:251). Tourism Web 2.0 applications, coined as "Travel 2.0" by Philip C. Wolf, the CEO of PhoCusWright, a successful consultancy firm, will evolve into Travel 3.0 technology with the development and use of Web 3.0. Travel 3.0 applications will generate predictions and recommendations for future travel experiences based on the past and current internet activities of travel consumers. Travel consumers provide invaluable information by posting comments and photos while planning trips online. In Travel 3.0, this historical data will be interpreted to develop personalized travel recommendations for travel consumers (Laboy & Torchio, 2007).

Web 2.0 or Travel 2.0 applications include content creation and management, interactive web application development techniques (AJAX), customer rating and evaluation systems, tagging, wikis, message boards, blogs, virtual worlds, and videos (Schmallegger & Carson, 2008). Thus, the effective use of these tools in the industry reveals customer preferences and needs, and the tourism sector adapts accordingly. For travel companies, having a website is not enough to reach potential customers on social media platforms. Travel companies should focus on social media platforms where the most suitable virtual tourism communities are located rather than maintaining a presence on all social media platforms. Search engines, bridging web sites and end-users, list any keyword users search for in their query, allowing them to find the information they are looking for more quickly. On TripAdvisor Consumer Review Site, tourists can add travel maps to the places they visit on the site. Tourist consumers can review and evaluate not only hotels but also all kinds of tourism-related products and services (Miguens et al., 2008). Another way to establish direct communication with customers in travel marketing is through email marketing. Email marketing is not about filling users' email inboxes with electronic mail but rather sending personalized emails. Sending such emails increases the return rates of customers (Kaur, 2017: 74). To keep customer relationships in the travel sector at the highest level positively, the most crucial sustainability issue in the tourism industry is the ability to respond to customers and their needs 24/7. Recognizing the problem of instantly meeting the demand for tourism, companies and their employees, in conjunction with the digitalization process, have introduced artificial intelligence. With the development of artificial intelligence, digital chatbots have emerged. These chatbots can guide people, provide responses to their requests, and offer support in the desired language 24/7. The active utilization of digital innovations such as reservation-taking, adjusting room temperature, lighting, and more in facilities allows quick responses to needs, thus improving customer satisfaction and quality.

## 1.4 Case Studies of Airlines and Tourism Companies Effectively Using Data to Personalize Services

Pegasus Airlines' mobile app allows passengers to search for flights, check-in, access mobile boarding passes, select seat numbers, and take advantage of various promotions. Passengers who purchase tickets through the mobile app earn twice as many points. If passengers need to cancel their flights for any reason and have more than 3 days before the departure, they can do so through the mobile app without incurring any fees (Pegasus Airlines, 2022). In 2018, a chatbot was implemented at the Swedish airport to provide services to passengers. These chatbots were integrated with Facebook Messenger, allowing passengers to obtain information about flights and airport services. Southwest Airlines introduced a chatbot to expedite and simplify their staff hiring processes, providing information to job applicants about open positions, skill requirements, and payment rates (Datasemantics, 2022). Lufthansa's mobile app offers passengers flight search, personal travel assistant, mobile check-in, boarding pass download, wallet, digital vaccine certificate, interactive airline app, and smart baggage features (Lufthansa.com, 2022). Atlanta's Hartsfield-Jackson International Airport (USA) features smart lanes, which aim to expedite the scanning process as it has 22 smart lanes in the security area, and it also includes other smart technologies like fingerprint scanning (Birkner, 2017). A fleet of autonomous baggage carts at Dallas Airport transported 450 bags per hour using half the energy of traditional baggage handling systems (Gindrat, 2021). At Heathrow Airport, cleaning robots are used to disinfect areas efficiently using ultraviolet (UV) light, which is shown to effectively kill harmful viruses and bacteria to ensure a safe and secure environment for passengers (Airport.nrdigital.com, 2022). Heathrow Airport's Positive Boarding project uses smart data tracking to improve flight punctuality and enhance the passenger experience, including the use of Smart Screen Technology and High-Tech Displays (NEC, 2017).

In South Korea, Seoul Incheon Airport introduced a robot named Troika, a large rectangular screen that resembles a giant smartphone and accompanies passengers to the departure gates, providing flight information, weather updates, and airport maps (Lee, 2017). At Dubai International Airport, passengers no longer need to remove their passports from their bags when passing through passport control. They can simply use their smartphones to scan their passports, and a variety of digital tools are implemented at the airport, including Smart Gates and Advanced Motion Sensor Systems (Mohamed et al., 2018). Turkish Airlines' mobile app allows passengers to purchase plane tickets, select menus, make reservations at accommodation facilities, choose preferred seats, request travel insurance, and access additional in-flight services (Turkishairlines.com, 2022). JetBlue Airlines opened the first fully integrated biometric self-boarding gate at JFK International Airport in New York (Corporatetraveller, 2022). Shenzhen Bao Airport in China uses the Anbot robot to perform security checks in the departure lounge of Terminal 3. Anbot, equipped with four high-resolution cameras and facial recognition software, can take photos of passengers and send them to security stations (Airport Robots Market, 2022). Similarly, airports like Rotterdam The Hague Airport, Gatwick, and Groupe ADP have experienced rapid growth in the use of autonomous vehicles in various operational areas, from baggage handling to cargo, with the Fleet innovation project featuring an autonomous baggage handling system (BHS) at Lahey Airport (Futuretravelexperience, 2018). Delta Air Lines has introduced smart baggage tags with RFID technology, allowing passengers to track the location of their bags through a smartphone app. RFID chips embedded in the baggage tags transmit information about the bag (including its geographical location) through radio waves detected by the airline's scanners, and this data is displayed on a map. Passengers can track the location of their bags by selecting the "Baggage Tracking" option in the Fly Delta app (Biersdorfer, 2017). Miami International Airport (MIA) has implemented a state-of-the-art automatic baggage system, which has doubled efficiency in baggage screening and delivery, at a cost of $324 million (International Airport Review, 2022).

## 2. CONCLUSION

The increasing capabilities of technological devices such as smartphones, tablets, and PCs, along with the decreasing production and usage costs, have facilitated the widespread adoption of digitalization, including Android and iOS software. One of the fastest-growing areas of digitalization is the tourism industry. Tourism businesses can more easily reach consumers in the virtual environment, and many businesses today make use of digital marketing. Businesses employ various techniques in digital marketing, such as SMS, email, MMS, social media, websites, and more. Industry 4.0, also referred to as digitalization, enables individuals to access information and compels businesses to incorporate numerous innovations into their business models in order to meet changing consumer expectations. The technological innovations brought about by Industry 4.0 create competition and opportunity advantages for the tourism sector. The opportunities arising from digital transformation will enable personalized services tailored to the preferences of tourists, making them feel special. The concept of "smart tourism" has transformed tourism products and services into advanced and different products and services through smart technologies. This transformation results in benefits such as reduced operating costs, reduced environmental impact, differentiation in services, ease of purchase for tourists, and competitive advantage in measuring digital tourism marketing investments.

The digital transformation and change that will affect all sub-sectors of tourism will inevitably impact aviation systems, which are directly related to technology. The aviation sector is one of the industries with the highest rate of technological turnover, and airports must adapt to these changes for sustainability. The scientific examination of digital transformations in aviation and airports is also of great importance. With this revolution, significant changes have already begun in airport systems, giving rise to the concept of "smart airports." Airports are major projects that influence the socioeconomic activities of the regions in which they are located. Therefore, airport investment plans are crucial for the efficient utilization of a country's resources. With digital transformation, it is anticipated that even more radical changes will occur in the aviation and airport sector in the future.

In today's world, where everything is automated and primarily relies on wireless technologies, the risk of networks being attacked or unauthorized users gaining access to and processing sensitive data is quite high. With technological advancements, threats have also increased. As known, in the context of an airport concept, this situation can mean unauthorized individuals gaining access to the private data of passengers and airport users. This, in turn, could pose a significant threat in the form of data breaches. In this scenario, uninvited guests can access and store data stored in airline databases, such as passengers' personal information or flight details. Furthermore, airports using the Internet of Things (IoT) have larger attack surfaces. Therefore, ensuring security is of paramount importance. Although such attacks may occur, the main goal is to minimize the impact on the entire network and affect as little functionality as possible. In this context, new studies should focus on data security and protection.

## REFERENCES


Airport Robots Market. (2022). *Airport Robots Market: New Innovations Trends, Research, Global Share and Growth Factor and Forecast: 2022-2031*. Benzinga. https://www.benzinga.com/pressreleases/22/04/26491728/airport-robots-marketnew-innovations-trends-research-global-share-and-growth-factor-and-forecast-2

Airport.nrdigital.com. (2022). *Robot Workforce: Airports Are Developing Robot Workers To Tirelessly Serve Passengers.* Airport NR Digital. https://airport.nridigital.com/air_may21/robot_airport_workers

Amadeus. (2017). *An Amadeus Company, Navitaire, Unveils The World's First.* Amadeus. https://amadeus.com/en/insights/press-release/an-amadeus-company-navitaireunveils-the-worlds-first-virtual-reality-travel-search-and-booking-experience

Angelakis, V., Tragos, E., & Pöhls, H. C. (2017). *Designing, Developing, And Facilitating Smart Cities*. Springer. doi:10.1007/978-3-319-44924-1

Atar, Y. (2019). *Turizm 4.0 (Dijital Dönüşüm).* Detay Yayıncılık.

Başaran, T., & Mercan, O. (2019). Sosyal Medyadaki Fotoğraf Ve Özçekimlerin Turizm Pazarlaması Açısından Önemi: İstanbul'a Yönelik Bir Çalışma. *1. International Congress on Dijital Transformation in Tourism I-DIGIT-19*: (pp. 115-128). IEEE.

Benyon, D., Quigley, A., O'Keefe, B., & Riva, G. (2013). Presence and dijital tourism. *AI & Society*, *29*(4), 521–529. doi:10.1007/s00146-013-0493-8

Biersdorfer, J. D. (2017). Luggage Tags That Help You Find Your Suitcase. *New York Times.* https://www.nytimes.com/2017/09/05/technology/personaltech/luggagetracking.html.

Bingemer, S. (2022). Global Distribution Systems (GDS). In *Encyclopedia of Tourism Management and Marketing*. Edward Elgar Publishing. doi:10.4337/9781800377486.global.distribution.systems

Birkner, C. (2017). How Delta's Focus on Customer Experience Turned. *AdWeek.* Http://Www.Adweek.Com/Brand-Marketing/How-Deltas-Focus-On-CustomerexperienceCreated-A-Powerhouse-Brand.

Boes, K. (2015). Smart Tourism Destinations: Smartness As Competitive Advantage. In ENTER 2015 PhD workshop research proposals, 11-15.

Budd, L., & Vorley, T. (2013). Airlines, Apps, And Business Travel: A Critical Examination. *Research in Transportation Business & Management*, *9*, 41–49. doi:10.1016/j.rtbm.2013.08.004

Buhalis, D., & Law, R. (2008). Progress in information technology and tourismmanagement: 20 years on and 10 yearsafterthe Internet—Thestate of eTourismresearch. *Tourism Management*, *29*(4), 609–623. doi:10.1016/j.tourman.2008.01.005

Chesher, M., Kaura, R., & Linton, P. (2013). Electronic Business And Commerce. Second Edition. Londra: Springer.

Corporatetraveller. (2022). *First Biometric Self-boarding At JFK.* Corporate Traveller. https://www.corporatetraveller.co.za/resources/in-the-press/jetblue-rolls-out-firstbiometric-self-boarding-at-jfk. 14.04.2022.

Datasemantics. (2022). *7 Chatbot Use Cases For Better Customer Experience At Airports.* Data Semantics. https://datasemantics.co/airport-chatbot/. 16.04.2022

Devries, P. (2008). The State Of RFID For E Ective Baggage Tracking İn The Airline Industry. *International Journal of Mobile Communications*, *6*(2), 151–164. doi:10.1504/IJMC.2008.016574

Dredge, D., Phi, G., Renuka, M., Eóin, M., & Popescu, E. S. (2018). Digitalisation in Tourism. European Commission's EASME Final Report. EC.

Esen, F., & Türkay, B. (2017). Turizm Endüstrilerinde Büyük Veri Kullanımı. *Journal of Tourism and Gastronomy Studies*, *5*(4), 92–115. doi:10.21325/jotags.2017.140

Fino, E. R., Martín-Gutiérrez, J., Fernández, M. D. M., & Davara, E. A. (2013). Interactive tourist guide: Connecting web 2.0, augmented reality and qr codes. *Procedia Computer Science*, *25*, 338–344. doi:10.1016/j.procs.2013.11.040

Gahi, Y., Guennoun, M., & Mouftah, H. T. (2016). *Big Data Analytics: Security and Privacy Challenges. 2016 IEEE Symposium on Computers and Communication (ISCC)*, Messina, Italy. 10.1109/ISCC.2016.7543859

Gindrat, R. (2021). Autonomous Vehicles For Airports: Multiple Use Cases. *Medium*. https://medium.com/bestmile/autonomous-vehicles-for-airports-multiple-use-cases12fbbeff3934. 14.04.2022.

Gretzel, U., Sigala, M., Xiang, Z., & Koo, C. (2015). Smart tourism: Foundations and developments. *Electronic Markets*, *25*(3), 179–188. doi:10.1007/s12525-015-0196-8

Güler, E. G. (2019). *Dijital dönüşüm ekseninde işletme uygulamaları içinde dijital dönüşüm ekseninde turizm uygulamaları*. Detay Yayıncılık.

Happ, E. and Horvath, Z. İ. (2018). Digital tourism is the challenge of future- a new approach to tourism. *Knowledge Horizons- Economics*, *10*(2), 9-16.

Heracleous, L., & Wirtz, J. (2006). Biometrics: The Next Frontier İn Service Excellence, Productivity And Security İn The Service Sector. *Managing Service Quality*, *16*(1), 12–22. hilton.com.tr/oteller/turkiye/istanbul/hilton-garden-inn-istanbul-golden-horn. doi:10.1108/09604520610639937

Howes, J. (2006). On The Cusp Of CUSS. *Kiosk Europe 18*. Spring 14. https://www.radissonhotels.com/tr-tr/oteller/radisson-blu-istanbul-tuzla-spa

International Air Transport Association. (2019). *IATA Global passenger survey highlights (IATA-GPS).* IATA. https://www.iata.org/en/publications

International Airport Review. (2022). *New Automated Baggage Handling System Unveiled At Miami Airport.* IAR. https://www.internationalairportreview.com/news/105683/new-automated-baggage-handling-system-miami-airport/

Jaffer, S., & Timbrell, G. (2014). Digital Strategy İn Airports. *25th Australasian Conference On Information System.* Australasian Conference On Information System.

Kaur, G. (2017). The Importance of Digital Marketing in the Tourism Industry. *International Journal of Research*, *5*, 72–77.

Koo, C., Gretzel, U., Hunter, W. C., & Chung, N. (2015). The Role of IT in Tourism. *Asia Pacific. Journal of Information Systems*, *25*(1), 99–104.

Kozak, Akoğlan, M., Evren, S. and Çakır, O. (2013). Tarihsel Süreç İçerisinde Turizm Paradigması. *Anatolia: Turizm Araştırmaları Dergisi*, *24*(1), 7–22.

Laboy, F., & Torchio, P. (2007). *Web 2.0 for the Travel Marketer and Consumer*. Saber Hospitality. https://www.sabrehospitality.com/pdf/whitepaper.pdf.

Law, R., Cheng Chu Chan, I., & Wang, L. (2018). A comprehensive review of mobile technology use in hospitality and tourism. *Journal of Hospitality Marketing & Management*, *27*(6), 626–648. doi:10.1080/19368623.2018.1423251

Lee, H., Chung, N., & Jung, T. (2015). Examining the cultural differences in acceptance of mobile augmented reality: Comparison of South Korea and Ireland. In *Information and communication technologies in tourism* (pp. 477–49). Springer.

Leung, D., Law, R., vanHoof, H., & Buhalis, D. (2013). Socialmedia in tourismandhospitality: A literaturereview. *Journal of Travel & Tourism Marketing*, *30*(1-2), 3–22. doi:10.1080/10548408.2013.750919

Li, J., Bon, M. A., & Ye, B. H. (2019). Hotel employee's artificial intelligence and robotics awareness and its impact on turnover intention: The moderating roles of perceived organizational support and competitive psychological climate. *Tourism Management*, *73*, 172–181. doi:10.1016/j.tourman.2019.02.006

Lufthansa.com. (2022). *Lufthansa apps*. Luftansa. https://www.lufthansa.com/rs/en/lufthansaapps. 16.04.2022.

Mariani, M. M., DiFelice, M. & Mura, M. (2016). Facebook as a destination marketing tool: Evidence from Italian regional Destination Management Organizations. *Tourism Management*, *54*, ss.321-343.

Miguéns, J., Baggio, R., & Costa, C. (2008). Social media and Tourism Destinations: TripAdvisor Case Study. *Advances in Tourism Research, Aveiro, Portugal*, (May), 26–28.

Mohamed, M. M., Gomaa, H. A. and El-Sherif, N. S. (2018). Evaluation of Current Smart Airport Technologies Implemented İn Cairo International Airport. *International Journal Of Heritage, Tourism And Hospitality*, *12* (2/2), 130-140.

NEC. (2017). *Smart Airport Supports Your Smooth Travel*. NEC. Www.Nec.Com

Pabel, A., & Prideaux, B. (2016). Social Media Use İn Pre-Trip Planning By Tourists Visiting A Small Regional Leisure Destination. *Journal of Vacation Marketing*, *22*(4), 335–348. doi:10.1177/1356766715618998

Polat, S., & Esen, F. (2021). Internet of Things For Travel Services. S. Polat & M. F. Esen (Eds.). Digital Technology Advancements In Knowledge Management. IGI Global.

Rawassizadeh, R., Price, B., & Petre, M. (2015). Wearables:Has The Age of Smartwatches Finally Arrived? *Communications of the ACM*, *58*(1), 45–47. doi:10.1145/2629633

Revfine. (2021). Alexa for hospitality; voice control speakers for hotels. *RevFine*. https://www.revfine.com/alexa-for-hospitality/.

Şanlı, S. C., Atçı, D., & Unur, K. (2019). Mersin'de Yiyecek İçecek İşletmelerinde Çalışan Personelin İşte Cep Telefonu Kullanımına İlişkin Algıları. *International Congress on "Dijital Transformation in Tourism I-DIGIT-19*, (pp. 26-42). IEEE.

Şanlıöz, K., Dilek, E. S., & Koçak, N. (2013). Değişen Dünya, Dönüşen Pazarlama: Türkiye Turizm Sektöründen Öncü Bir Mobil Uygulama Örneği. *Anatolia: Turizm Araştırmaları Dergisi*, *24*(2), 250–260.

Schmallegger, D., & Carson, D. (2008). Blogs in Tourism: Changing Approaches to Information Exchange. *Journal of Vacation Marketing*, *14*(2), 99–110. doi:10.1177/1356766707087519

Song, H., & Liu, H. (2017). Predicting tourist demand using big data. In *Analytics in smart tourism design* (pp. 13–29). Springer. doi:10.1007/978-3-319-44263-1_2

Türsab (2020). *Türkiye ve Dünya Turizmi Değerlendirmesi*. Tursab.

Türsab & TBV. (2020). Turizm Sektörü Dijitalleşme Yol Haritası. Tursab. https://www.tursab.org.tr/apps/Files/Content/ad5f3ddb-5a11-410f-9e3cfe8b2dc4df8b.pdf,.

Urry, J. (1999). *Mekânları Tüketmek*. Mart Yayıncılık.

Wang, D. & ve Xiang, Z. (2012). The New Landscape of Travel: A Comprehensive Analysis of Smartphone Apps. In M. Fuchs, F. Ricci ve L. Cantoni (Eds.), Information and Communication Technologies in Tourism. Wien: Springer.

Want, R. (2006). An Introduction To RFID Technology. *IEEE Pervasive Computing*, *5*(1), 25–33. doi:10.1109/MPRV.2006.2

Wittmer, A. (2011). Acceptance Of Self-Service Check-İn At Zurich Airport. *Research in Transportation Business & Management*, *1*(1), 136–143. doi:10.1016/j.rtbm.2011.06.001

Wu, W. (2020). *Analysis of Digital Tourism, Virtual Tourism and Wisdom Tourism*. Springer. https://link.springer.com/conference/csia

Yalçın Kayıkçı, M., & Kutluk Bozkurt, A. (2018). Dijital Çağda Z Ve Alpha Kuşağı, Yapay Zeka Uygulamaları Ve Turizme Yansımaları. *Sosyal Bilimler Metinleri*, *01*, 54–64.

# Conclusion

In the ever-evolving landscape of air transportation, *Harnessing Digital Innovation for Air Transportation* emerges not only as a comprehensive guide but as a testament to the transformative power of digitalization in revolutionizing the aviation industry. As we conclude our journey through the rich tapestry of insights woven throughout this book, it becomes increasingly evident that the impact of digital innovation extends far beyond the confines of individual chapters—it permeates every facet of aviation operations, shaping the future of air travel in profound ways.

Throughout the book, readers have been invited to explore the dynamic intersection of technology, strategy, and innovation within the aviation sector. From the pioneering applications of digital assistants to the strategic integration of artificial intelligence and data analytics, each chapter offers a nuanced understanding of how digitalization is reshaping traditional paradigms and driving unprecedented advancements in safety, efficiency, and passenger experience.

Yet, as we bid farewell to these pages, it's essential to recognize that the journey towards digital transformation in air transportation is far from over. While the book provides a comprehensive overview of the current landscape, it also serves as a catalyst for further exploration and discovery. By addressing lingering issues and closing open loops, *Harnessing Digital Innovation for Air Transportation* empowers readers to embark on their own quest for knowledge, innovation, and excellence in the aviation industry.

Looking ahead, the conclusion offers a compelling call to action—a call to embrace digital innovation with unwavering enthusiasm and commitment. Whether you are a seasoned industry veteran, a curious researcher, or an aspiring aviation professional, there is much to be done. By fostering collaboration, driving innovation, and championing digital transformation, each reader has the opportunity to play a pivotal role in shaping the future of air transportation.

Moreover, in the spirit of providing even more value, readers are encouraged to leverage additional resources and avenues for learning. Whether it's engaging with industry experts, participating in professional development opportunities, or staying abreast of the latest technological advancements, the journey towards digital innovation in air transportation is one of continuous growth and exploration.

In closing, *Harnessing Digital Innovation for Air Transportation* stands as a beacon of inspiration and insight—a testament to the boundless potential of digitalization to propel the aviation industry forward into a new era of progress and possibility. As readers turn the final page and embark on their own journeys, may the lessons learned and insights gained from this book serve as a guiding light, illuminating the path towards a future where innovation knows no bounds and the sky truly is the limit.

# Compilation of References


Abele-Wigert, I., & Dunn, M. (2006). The International CIIP Handbook 2006: An Inventory of Protection Policies in 20 Countries and 6 [). Center for Security Studies.]. *International Organization, I.*

Agyabeng-Mensah, Y., Ahenkorah, E., Afum, E., Agyemang, A. N., Agnikpe, C., & Rogers, F. (2020). Examining the influence of internal green supply chain practices, green human resource management and supply chain environmental cooperation on firm performance. *Supply Chain Management*, *25*(5), 585–599. doi:10.1108/SCM-11-2019-0405

Airport Robots Market. (2022). *Airport Robots Market: New Innovations Trends, Research, Global Share and Growth Factor and Forecast: 2022-2031*. Benzinga. https://www.benzinga.com/pressreleases/22/04/26491728/airport-robots-marketnew-innovations-trends-research-global-share-and-growth-factor-and-forecast-2

Airport.nrdigital.com. (2022). *Robot Workforce: Airports Are Developing Robot Workers To Tirelessly Serve Passengers.* Airport NR Digital. https://airport.nridigital.com/air_may21/robot_airport_workers

Airport-Technology.com. (2021, February 9). *Biggest influencers in Airlines in Q4 2020: Top individuals and companies to follow.* Airport Technology. https://www.airport-technology.com/dashboards/influencers-dashboards/biggest-influencers-airlines/

Alam, M. A. (2016). Techno-stress and productivity: Survey evidence from the aviation industry. *Journal of Air Transport Management*, *50*, 62–70. doi:10.1016/j.jairtraman.2015.10.003

Allianz.com.tr. (2020). *Günümüzün Popüler Mesleği Influencer Nedir? Influencer Nasıl Olunur?* Allianz. https://www.allianz.com.tr/tr_TR/seninle-guzel/gunumuzun-populer-meslegi-influencer-nedir.html

Alraies, M. C., & Sahni, S. (2017). Why cardiologists should be on social media–the value of online engagement. *Expert Review of Cardiovascular Therapy*, *15*(12), 889–890. doi:10.1080/14779072.2017.1408408 PMID:29164942

Amadeus. (2017). *An Amadeus Company, Navitaire, Unveils The World's First.* Amadeus. https://amadeus.com/en/insights/press-release/an-amadeus-company-navitaireunveils-the-worlds-first-virtual-reality-travel-search-and-booking-experience

Angelakis, V., Tragos, E., & Pöhls, H. C. (2017). *Designing, Developing, And Facilitating Smart Cities*. Springer. doi:10.1007/978-3-319-44924-1

Arjomandi, M., Agostino, S., Mammone, M., Nelson, M., & Zhou, T. (2006). Classification of unmanned aerial vehicles. Report for Mechanical Engineering class, University of Adelaide, Adelaide, Australia, 1-48.

Asbaş, C., & Erdem Tuzlukaya, Ş. (2022). Cyberattack and cyberwarfare strategies for businesses. In F. Özsungur (Ed.), *Conflict Management in Digital Business: New Strategy and Approach* (pp. 303–329). Emerald Publishing. doi:10.1108/978-1-80262-773-220221027

Asbaş, C., & Erdem Tuzlukaya, Ş. (2023). Cyberwarfare: War activities in cyberspace. In F. Özsungur (Ed.), *Handbook of Research on War Policies, Strategies, and Cyber Wars* (pp. 128–145). IGI Global. doi:10.4018/978-1-6684-6741-1.ch007

Aslan, M. E., & Tolga, A. C. (2022). Evaluation of Artificial Intelligence Applications in Aviation Maintenance, Repair and Overhaul Industry via MCDM Methods. In C. Kahraman, A. C. Tolga, S. Cevik Onar, S. Cebi, B. Oztaysi, & I. U. Sari (Eds.), *Intelligent and Fuzzy Systems. INFUS 2022. Lecture Notes in Networks and Systems* (Vol. 504). Springer., doi:10.1007/978-3-031-09173-5_94

ATAG. (2020). *Facts.* Aviation Benefits Beyond Borders. https://atag.org/facts-figures/

Atar, Y. (2019). *Turizm 4.0 (Dijital Dönüşüm).* Detay Yayıncılık.

Avram, B. (2017). Ancillaries in the aviation industry. Importance, trends, going digital. *Expert Journal of Marketing*, *5*(2), 53–65.

Bachtiar, T. A. (2009). *Hamas: Kenapa dibenci Israel?* Hikma.

Bandyopadhyay, T., Sen, S., & Dutta, S. (2018). Artificial Intelligence in Air Traffic Management Systems: A Review. *2018 International Conference on Computing, Power and Communication Technologies (GUCON).*

Barbe, D., & Neuburger, L. (2021). Generation Z and Digital Influencers in the Tourism Industry. In N. Stylos, R. Rahimi, B. Okumus, & S. Williams (Eds.), *Generation Z Marketing and Management in Tourism and Hospitality* (pp. 167–192). Palgrave Macmillan. doi:10.1007/978-3-030-70695-1_7

Barnhart, R. K., Hottman, S. B., Marshall, D. M., & Shappee, E. (2012). Introduction to Unmanned Aircraft Systems. CRC Press, Taylor & Francis Group, FL, USA.

Başaran, T., & Mercan, O. (2019). Sosyal Medyadaki Fotoğraf Ve Özçekimlerin Turizm Pazarlaması Açısından Önemi: İstanbul'a Yönelik Bir Çalışma. *1. International Congress on Dijital Transformation in Tourism I-DIGIT-19*: (pp. 115-128). IEEE.

Batouei, A., Iranmanesh, M., Nikbin, D., & Hyun, S. S. (2019). Flight anxiety: Investigating the role of airline service quality and flight crew's competence. *Asia Pacific Journal of Tourism Research*, *24*(7), 710–724. doi:10.1080/10941665.2019.1630457

Belhadi, A., Kamble, S., Jabbour, C. J. C., Gunasekaran, A., Ndubisi, N. O., & Venkatesh, M. (2021). Manufacturing and service supply chain resilience to the COVID-19 outbreak: Lessons learned from the automobile and airline industries. *Technological Forecasting and Social Change*, *163*, 120447. doi:10.1016/j.techfore.2020.120447 PMID:33518818

Benbya, H., Davenport, T. H., & Pachidi, S. (2020). Artificial Intelligence in Organizations: Current State and Future Opportunities (December 3, 2020). [Available at SSRN: https://ssrn.com/abstract=3741983 or http://dx.doi.org/10.2139/ssrn.3741983]. *MIS Quarterly Executive*, *19*(4), 4. Advance online publication.

Benyon, D., Quigley, A., O'Keefe, B., & Riva, G. (2013). Presence and dijital tourism. *AI & Society*, *29*(4), 521–529. doi:10.1007/s00146-013-0493-8

Berman, S. J. (2012). Digital transformation: Opportunities to create new business models. *Strategy and Leadership*, *40*(2), 16–24. doi:10.1108/10878571211209314

Bhatti, G., Mohan, H., & Singh, R. R. (2021). Towards the future of smart electric vehicles: Digital twin technology. *Renewable & Sustainable Energy Reviews*, *141*, 110801. doi:10.1016/j.rser.2021.110801

Biersdorfer, J. D. (2017). Luggage Tags That Help You Find Your Suitcase. *New York Times.* https://www.nytimes.com/2017/09/05/technology/personaltech/luggagetracking.html.

Bigne, E., Andreu, L., Hernandez, B., & Ruiz, C. (2018). The impact of social media and offline influences on consumer behaviour. An analysis of the low-cost airline industry. *Current Issues in Tourism*, *21*(9), 1014–1032. doi:10.1080/13683500.2015.1126236

Bingemer, S. (2022). Global Distribution Systems (GDS). In *Encyclopedia of Tourism Management and Marketing*. Edward Elgar Publishing. doi:10.4337/9781800377486.global.distribution.systems

Binthiyaa, B., & Fazıl, A. (2023). Impact of social media in airline industry. *Eur. Chem. Bull.* *12*(4), 9141-9148. https://www.eurchembull.com/uploads/paper/80a0722b8ddad491b0c6d6550557eeba.pdf

Birkner, C. (2017). How Delta's Focus on Customer Experience Turned. *AdWeek*. Http://Www.Adweek.Com/Brand-Marketing/How-Deltas-Focus-On-CustomerexperienceCreated-A-Powerhouse-Brand.

Blom, J. D. (2010). *Unmanned Aerial Systems: A Historical Perspective*. Combat Studies Institute Press.

Boes, K. (2015). Smart Tourism Destinations: Smartness As Competitive Advantage. In ENTER 2015 PhD workshop research proposals, 11-15.

Bozkuş̧ Kahyaoğlu, S., & Çalıyurt, K. (2018). Cyber security assurance process from the internal audit perspective. *Managerial Auditing Journal*, *33*(4), 360–376. doi:10.1108/MAJ-02-2018-1804

Brett, D. (2019) Unilode and OnAsset complete Bluetooth ULD trial. AirCargoNews. https://www.aircargonews.net/airlines/unilode-and-onasset-completebluetooth-uld-trial/

Brodie, R. J., Hollebeek, L. D., Jurić, B., & Ilić, A. (2011). Customer engagement: Conceptual domain, fundamental propositions, and implications for research. *Journal of Service Research*, *14*(3), 252–271. doi:10.1177/1094670511411703

Broekel, T., & Boschma, R. (2012). Knowledge networks in the Dutch aviation industry: The proximity paradox. *Journal of Economic Geography*, *12*(2), 409–433. doi:10.1093/jeg/lbr010

Budd, L., & Vorley, T. (2013). Airlines, Apps, And Business Travel: A Critical Examination. *Research in Transportation Business & Management*, *9*, 41–49. doi:10.1016/j.rtbm.2013.08.004

Buhalis, D., & Law, R. (2008). Progress in information technology and tourismmanagement: 20 years on and 10 yearsafterthe Internet—Thestate of eTourismresearch. *Tourism Management*, *29*(4), 609–623. doi:10.1016/j.tourman.2008.01.005

Buss, A., Pop, C. L., Reid, G., & Kirsch, A. (2020). User Experience in Unmanned Aircraft Systems: A Survey and Research Agenda. *IEEE Access : Practical Innovations, Open Solutions*, *8*, 144421–144434.

Campbell, C., & Farrell, J. R. (2020). More than meets the eye: The functional components underlying influencer marketing. *Business Horizons*, *63*(4), 469–479. doi:10.1016/j.bushor.2020.03.003

Cavelty, M. D. (2015). Cyber-security and private actors. In R. Abrahamsen & A. Leander (Eds.), *Routledge Handbook of Private Security Studies* (pp. 89–99). Routledge. doi:10.4324/9781315850986-10

Chan, A. P. C., Yung, E. H. K., Lam, P. T. I., Tam, C. M., & Cheung, S. O. (2001). Application of Delphi method in selection of procurement systems for construction projects. *Construction Management and Economics*, *19*(7), 699–718. doi:10.1080/01446190110066128

Cheng, T., Wen, P., & Li, Y. (2016, December). Research status of artificial neural network and its application assumption in aviation. In 2016 12th International Conference on Computational Intelligence and Security (CIS) (pp. 407-410). IEEE.

Chesher, M., Kaura, R., & Linton, P. (2013). Electronic Business And Commerce. Second Edition. Londra: Springer.

Chiaramonti, D. (2019). Sustainable aviation fuels: The challenge of decarbonization. *Energy Procedia*, *158*, 1202–1207. doi:10.1016/j.egypro.2019.01.308

Chumwatana, T., & Chuaychoo, I. (2017, October). Using social media listening technique for monitoring people's mentions from social media: A case study of Thai airline industry. In *2017 2nd International Conference on Knowledge Engineering and Applications (ICKEA)* (pp. 103-106). IEEE. 10.1109/ICKEA.2017.8169910

Cichonski, P., Millar, T., Grance, T., & Scarfone, K. (2012). *Computer Security Incident Handling Guide: Recommendations of the National Institute of Standards and Technology*. NIST. doi:10.6028/NIST.SP.800-61r2

Cir, I. C. A. O. 328 (2011). Unmanned Aircraft Systems. Springer.

Colbert, A., Yee, N., & George, G. (2016). The digital workforce and the workplace of the future. *Academy of Management Journal, 59*(3), 731–739. doi:10.5465/amj.2016.4003

Coldwell, D. A. (2019). Negative influences of the 4th industrial revolution on the workplace: Towards a theoretical model of entropic citizen behavior in toxic organizations. *International Journal of Environmental Research and Public Health, 16*(15), 2670. doi:10.3390/ijerph16152670 PMID:31349722

Coletta, P. E. (1987). *Patrick N. L. Bellinger and U.S. Naval Aviation.* University Press of America, Inc.

Conde, R., & Casais, B. (2023). Micro, macro and mega-influencers on instagram: The power of persuasion via the parasocial relationship. *Journal of Business Research, 158*, 113708. doi:10.1016/j.jbusres.2023.113708

Condron, S. (2007). Getting it right: Protecting American critical infrastructure in cyberspace. *Harvard Journal of Law & Technology, 20*(2), 403–422.

Corbin, J., & Strauss, A. (2008). *Basics of Qualitative Research: Techniques and Procedures for Developing Grounded Theory* (3rd ed.). Sage Publications. doi:10.4135/9781452230153

Cornish, P. N., Hughes, R., & Livingstone, D. (2009). *Cyberspace and the National Security of the United Kingdom: Threats and Responses.* Chatham House.

Corporatetraveller. (2022). *First Biometric Self-boarding At JFK.* Corporate Traveller. https://www.corporatetraveller.co.za/resources/in-the-press/jetblue-rolls-out-firstbiometric-self-boarding-at-jfk. 14.04.2022.

Datasemantics. (2022). *7 Chatbot Use Cases For Better Customer Experience At Airports.* Data Semantics. https://datasemantics.co/airport-chatbot/. 16.04.2022

Dazhi, H. (2017). *Post-evaluation research on the oil supply project of Yantai Chaoshui Airport* [Dissertation, Qingdao University], Qingdao, China.

Deibert, R. J., & Rohozinski, R. (2010). Risking security: Policies and paradoxes of cyberspace security. *International Political Sociology, 4*(1), 15–32. doi:10.1111/j.1749-5687.2009.00088.x

Dessart, L. (2017). Social media engagement: A model of antecedents and relational outcomes. *Journal of Marketing Management, 33*(5-6), 375–399. doi:10.1080/0267257X.2017.1302975

Dessart, L., Veloutsou, C., & Morgan-Thomas, A. (2015). Consumer engagement in online brand communities: A social media perspective. *Journal of Product and Brand Management, 24*(1), 28–42. doi:10.1108/JPBM-06-2014-0635

Dessens, O., Köhler, M. O., Rogers, H. L., Jones, R. L., & Pyle, J. A. (2014). Aviation and climate change. *Transport Policy, 34*, 14–20. doi:10.1016/j.tranpol.2014.02.014

Devlin, J., Chang, M. W., Lee, K., & Toutanova, K. (2018). *BERT: Pre-training of deep bidirectional transformers for language understanding*. In *Proceedings of NAACL-HLT 2019*, Minneapolis, MN, US.

Devries, P. (2008). The State Of RFID For E Ective Baggage Tracking İn The Airline Industry. *International Journal of Mobile Communications*, *6*(2), 151–164. doi:10.1504/IJMC.2008.016574

Digitalage.com.tr. (2023, March 9). *Markaların Influencer Marketing Yatırımlarının En Büyük Engeli Samimiyet.*. Digital Age. https://digitalage.com.tr/markalarin-influencer-marketing-yatirimlarinin-en-buyuk-engeli-samimiyet/

Dijkmans, C., Kerkhof, P., & Beukeboom, C. J. (2015). A stage to engage: Social media use and corporate reputation. *Tourism Management*, *47*, 58–67. doi:10.1016/j.tourman.2014.09.005

Dožić, S. (2019). Multi-criteria decision making methods: Application in the aviation industry. *Journal of Air Transport Management*, *79*, 101683. doi:10.1016/j.jairtraman.2019.101683

Dredge, D., Phi, G., Renuka, M., Eóin, M., & Popescu, E. S. (2018). Digitalisation in Tourism. European Commission's EASME Final Report. EC.

Dube, K., Nhamo, G., & Chikodzi, D. (2021). COVID-19 pandemic and prospects for recovery of the global aviation industry. *Journal of Air Transport Management*, *92*, 102022. doi:10.1016/j.jairtraman.2021.102022 PMID:36567961

Dyson, E., Gilder, G., Keyworth, G., & Toffler, A. (1996). Cyberspace and the American dream: A Magna Carta for the knowledge age. *The Information Society*, *12*(3), 295–308. doi:10.1080/019722496129486

EASA. (2023). *Urban Air Mobility (UAM)*. EASA. https://www.easa.europa.eu/en/light/topics/urban-air-mobility-uam

Elhmoud, E. R., & Kutty, A. A. (2021). Sustainability assessment in aviation industry: A mini-review on the tools, models and methods of assessment. In *Proceedings of the 2nd African International Conference on Industrial Engineering and Operations Management, Harare, Zimbabwe* (pp. 7-10). IEOM Society. http://ieomsociety.org/harare2020/papers/324.pdf

Ellrich, F., Bauer, M., Schreiner, N., Keil, A., Pfeiffer, T., Klier, J., Weber, S., Jonuscheit, J., Friederich, F., & Molter, D. (2020). Terahertz quality inspection for automotive and aviation industries. *Journal of Infrared, Millimeter and Terahertz Waves*, *41*(4), 470–489. doi:10.1007/s10762-019-00639-4

Esen, F., & Türkay, B. (2017). Turizm Endüstrilerinde Büyük Veri Kullanımı. *Journal of Tourism and Gastronomy Studies*, *5*(4), 92–115. doi:10.21325/jotags.2017.140

European Union Aviation Safety Agency. EASA, (2020). Artificial Intelligence Roadmap: A human-centric approach to AI in Aviation. Retrieved October 12, 2023 from: http://www.adfingo.com/easa-ai-roadmap-v10.pdf

Even, S., & Siman-Tov, D. (2012). *Cyber Warfare: Concepts and Strategic Trends*. Institute for National Security Studies.

FAA (2023). *Urban Air Mobility (UAM) Version 2.0-Concept of Operations*. FAA.

Fahlstrom, P. G., & Gleason, T. J. (2012). *Introduction to UAV Systems* (4th ed.). John Wiley & Sons, Ltd.

Fazıl, A. (2023). Impact Of Socıal Medıa In Aırlıne Industry. *Eur. Chem. Bull. 12*(4), 9141-9148.

Feedspot.com. (2023, August 7). *Top 45 Aviation Influencers in 2023*. FeedSpot. https://blog.feedspot.com/aviation_instagram_influencers/

Fino, E. R., Martín-Gutiérrez, J., Fernández, M. D. M., & Davara, E. A. (2013). Interactive tourist guide: Connecting web 2.0, augmented reality and qr codes. *Procedia Computer Science, 25*, 338–344. doi:10.1016/j.procs.2013.11.040

Frenken, K. (2000). A complexity approach to innovation networks. The case of the aircraft industry (1909–1997). *Research Policy, 29*(2), 257–272. doi:10.1016/S0048-7333(99)00064-5

Fritz, S., See, L., Carlson, T., Haklay, M., Oliver, J. L., Fraisl, D., Mondardini, R., Brocklehurst, M., Shanley, L. A., Schade, S., Wehn, U., Abrate, T., Anstee, J., Arnold, S., Billot, M., Campbell, J., Espey, J., Gold, M., Hager, G., & West, S. (2019). Citizen science and the United Nations sustainable development goals. *Nature Sustainability, 2*(10), 922–930. doi:10.1038/s41893-019-0390-3

Gahi, Y., Guennoun, M., & Mouftah, H. T. (2016). *Big Data Analytics: Security and Privacy Challenges. 2016 IEEE Symposium on Computers and Communication (ISCC)*, Messina, Italy. 10.1109/ISCC.2016.7543859

Ghadiminia, N., Mayouf, M., Cox, S., & Krasniewicz, J. (2021). BIM-enabled facilities management (FM): A scrutiny of risks resulting from cyber attacks. *Journal of Facilities Management*. doi:10.1108/JFM-01-2021-0001

Gill, G. K., & Shergill, G. S. (2004). Perceptions of safety management and safety culture in the aviation industry in New Zealand. *Journal of Air Transport Management, 10*(4), 233–239. doi:10.1016/j.jairtraman.2004.02.002

Gindrat, R. (2021). Autonomous Vehicles For Airports: Multiple Use Cases. *Medium*. https://medium.com/bestmile/autonomous-vehicles-for-airports-multiple-use-cases12fbbeff3934. 14.04.2022.

Gisario, A., Kazarian, M., Martina, F., & Mehrpouya, M. (2019). Metal additive manufacturing in the commercial aviation industry: A review. *Journal of Manufacturing Systems, 53*, 124–149. doi:10.1016/j.jmsy.2019.08.005

Gold, H., Faqiri, H., Yeung, J., & Hu, C. (2023, October 8). Israel formally declares war against Hamas as it battles to push militants off its soil. CNN. Retrieved from https://www.cnn.com/2023/10/08/middleeast/israel-gaza-attack-hostages-response-intl-hnk/index.html

Gonzalez-Garay, A., Heuberger-Austin, C., Fu, X., Klokkenburg, M., Zhang, D., van der Made, A., & Shah, N. (2022). Unravelling the potential of sustainable aviation fuels to decarbonise the aviation sector. *Energy & Environmental Science, 15*(8), 3291–3309. doi:10.1039/D1EE03437E

Gretzel, U., Sigala, M., Xiang, Z., & Koo, C. (2015). Smart tourism: Foundations and developments. *Electronic Markets, 25*(3), 179–188. doi:10.1007/s12525-015-0196-8

Grieves, M. (2015). *Digital twin: Manufacturing excellence through virtual factory replication.* (Digital twin white paper). FIT. https://research.fit.edu/media/sitespecific/researchfitedu/camid/documents/1411.0_Digital_Twin_White_ Paper_Dr_Grieves.pdf

Güler, E. G. (2019). *Dijital dönüşüm ekseninde işletme uygulamaları içinde dijital dönüşüm ekseninde turizm uygulamaları.* Detay Yayıncılık.

Gupta, S. J., Ghonge, M. M., & Jawandhiya, P. M. (2013). Review of Unmanned Aircraft System (UAS). [IJARCET]. *International Journal of Advanced Research in Computer Engineering and Technology, 2*(4), 1646–1658.

Happ, E. and Horvath, Z. İ. (2018). Digital tourism is the challenge of future- a new approach to tourism. *Knowledge Horizons- Economics, 10*(2), 9-16.

Hassanalian, M., & Abdelkef, A. (2017). Classifications, applications, and design challenges of drones: A review. *Progress in Aerospace Sciences, 91*, 99–131. doi:10.1016/j.paerosci.2017.04.003

Hathaway, O. A., Crootof, R., Levitz, P., Nix, H., Nowlan, A., Perdue, W., & Spiegel, J. (2012). The Law of Cyber-Attack. *California Law Review, 100*(4), 817–885.

Hays, S., Page, S. J., & Buhalis, D. (2013). Social media as a destination marketing tool: Its use by national tourism organisations. *Current Issues in Tourism, 16*(3), 211–239. doi:10.1080/13683500.2012.662215

Heracleous, L., & Wirtz, J. (2006). Biometrics: The Next Frontier İn Service Excellence, Productivity And Security İn The Service Sector. *Managing Service Quality, 16*(1), 12–22. hilton.com.tr/oteller/turkiye/istanbul/hilton-garden-inn-istanbul-golden-horn. doi:10.1108/09604520610639937

Hook, L. R., Sizoo, D., & Fuller, J. G. (2022). *How digital safety systems could revolutionize aviation safety. In 2022 IEEE/AIAA 41st Digital Avionics Systems Conference (DASC).* IEEE. https://www.iata.org/en/events/all/world-sustainability,

Howes, J. (2006). On The Cusp Of CUSS. *Kiosk Europe 18.* Spring 14. https://www.radissonhotels.com/tr-tr/oteller/radisson-blu-istanbul-tuzla-spa

Iasiello, E. (2014). A review of "a framework for aviation cybersecurity" – a decision paper from the American Institute of Aeronautics and Astronautics. *Journal of Homeland Security and Emergency Management, 11*(1), 95–100. doi:10.1515/jhsem-2013-0076

IATA. (2023). Digital Innovation Program. IATA. https://www.iata.org/en/programs/innovation/digital-innovation/

IATA. (2023). World Sustainability Symposium (WSS). IATA. https://www.iata.org/en/events/all/world-sustainability-symposium/

Ibrion, M., Paltrinieri, N., & Nejad, A. R. (2019). On risk of digital twin implementation in marine industry: Learning from aviation industry. *Journal of Physics, 1357*(1), 012009.

Ibrion, M., Paltrinieri, N., & Nejad, A. R. (2019, October). On risk of digital twin implementation in marine industry: Learning from aviation industry. [). IOP Publishing.]. *Journal of Physics: Conference Series, 1*(1), 012009. doi:10.1088/1742-6596/1357/1/012009

Influencermarketing.ai. (2023). *Top Travel, Tourism, And Aviation Instagram Influencers To Follow In Celle For 2023.* Influence Marketing. https://influencermarketing.ai/top-travel-tourism-and-aviation-instagram-influencers-to-follow-in-celle-for-2023/

International Air Transport Association. (2019). *IATA Global passenger survey highlights (IATA-GPS).* IATA. https://www.iata.org/en/publications

International Airport Review. (2022). *New Automated Baggage Handling System Unveiled At Miami Airport.* IAR. https://www.internationalairportreview.com/news/105683/new-automated-baggage-handling-system-miami-airport/

Jaffer, S., & Timbrell, G. (2014). Digital Strategy İn Airports. *25th Australasian Conference On Information System.* Australasian Conference On Information System.

Jansen, C., & Jeschke, S. (2018). Mitigating risks of digitalization through managed industrial security services. *AI & Society, 33*(2), 163–173. doi:10.1007/s00146-018-0812-1

Javornik, M., Nadoh, N., & Lange, D. (2019). Data is the new oil. In B. Müller & G. Meyer (Eds.), *Towards User-Centric Transport in Europe. Lecture Notes in Mobility.* Springer. doi:10.1007/978-3-319-99756-8_19

Kang, M., & Schuett, M. A. (2013). Determinants of sharing travel experiences in social media. *Journal of Travel & Tourism Marketing, 30*(1-2), 93–107. doi:10.1080/10548408.2013.751237

Karabacak, B., Yıldırım, S. Ö., & Baykal, N. (2016). A vulnerability-driven cyber security maturity model for measuring national critical infrastructure protection preparedness. *International Journal of Critical Infrastructure Protection, 15*, 47–59. doi:10.1016/j.ijcip.2016.10.001

Karaman, A. S., Kilic, M., & Uyar, A. (2018). Sustainability reporting in the aviation industry: Worldwide evidence. *Sustainability Accounting. Management and Policy Journal, 9*(4), 362–391.

Kashyap, R. (2019). Artificial Intelligence Systems in Aviation. In T. Shmelova, Y. Sikirda, N. Rizun, & D. Kucherov (Eds.), *Cases on Modern Computer Systems in Aviation* (pp. 1–26). IGI Global., doi:10.4018/978-1-5225-7588-7.ch001

Kaur, G. (2017). The Importance of Digital Marketing in the Tourism Industry. *International Journal of Research, 5*, 72–77.

Keane, J. F., & Carr, S. S. (2013). A Brief History of Early Unmanned Aircraft. *Johns Hopkins APL Technical Digest*, *32*(3), 558–571.

Kim, B., Yoo, M., & Yang, W. (2020). Online engagement among restaurant customers: The importance of enhancing flow for social media users. *Journal of Hospitality & Tourism Research (Washington, D.C.)*, *44*(2), 252–277. doi:10.1177/1096348019887202

Knoblich, S., Martin, A., Nash, R., & Stansbie, P. (2017). Keys to success in Social Media Marketing (SMM)–Prospects for the German airline industry. *Tourism and Hospitality Research*, *17*(2), 147–164. doi:10.1177/1467358415610415

Kok, S. L., & Siripipatthanakul, S. (2023). Change Management Model in Corporate Culture and Values: A Case Study of Intel Cooperation. *Advance Knowledge for Executives*, *2*(1), 1–30.

Koo, C., Gretzel, U., Hunter, W. C., & Chung, N. (2015). The Role of IT in Tourism. *Asia Pacific. Journal of Information Systems*, *25*(1), 99–104.

Korchenko, A. G. & ve Illyash, O.S. (2013). The Generalized Classification of Unmanned Air Vehicles. *IEEE 2nd International Conference Actual Problems of Unmanned Air Vehicles Developments*. IEEE.

Koroniotis, N., Moustafa, N., Schiliro, F., Gauravaram, P., & Janicke, H. (2020). A holistic review of cybersecurity and reliability perspectives in smart airports. *IEEE Access : Practical Innovations, Open Solutions*, *8*, 209802–209834. doi:10.1109/ACCESS.2020.3036728

Kosuda, M., Lipovsky, P., Yang, W., Bai, M., Novotnak, J., & Szoke, Z. (2020). *Discussion on blockchain applications in unmanned aerial systems domain. 2020 New Trends in Aviation Development*. NTAD., doi:10.1109/NTAD51447.2020.9379109

Kozak, Akoğlan, M., Evren, S. and Çakır, O. (2013). Tarihsel Süreç İçerisinde Turizm Paradigması. *Anatolia: Turizm Araştırmaları Dergisi*, *24*(1), 7–22.

Kulida, E. L., & Lebedev, V. G. (2020). About the Use of Artificial Intelligence Methods in Aviation. 2020 13th International Conference "Management of large-scale system development" (MLSD), Moscow, Russia, 2020, pp. 1-5, doi: 10.1109/MLSD49919.2020.9247822.1-5

Kumar, M. (2022). Optimized application of artificial intelligence (AI) in aviation market. *International Journal of Recent Research Aspects*, *9*(4).

Laboy, F., & Torchio, P. (2007*). Web 2.0 for the Travel Marketer and Consumer*. Saber Hospitality. https://www.sabrehospitality.com/pdf/whitepaper.pdf.

Lanshina, T. A., Barinova, V. A., Kondratyev, A. D., & Romantsov, M. V. (2020). Sustainable development and digitalization: The unusual COVID-19 crisis requires original solutions. *Bulletin of International Organizations*, *15*(4), 91–114. doi:10.17323/1996-7845-2020-04-05

Law, R., Cheng Chu Chan, I., & Wang, L. (2018). A comprehensive review of mobile technology use in hospitality and tourism. *Journal of Hospitality Marketing & Management*, *27*(6), 626–648. doi:10.1080/19368623.2018.1423251

Lee, H., Chung, N., & Jung, T. (2015). Examining the cultural differences in acceptance of mobile augmented reality: Comparison of South Korea and Ireland. In *Information and communication technologies in tourism* (pp. 477–49). Springer.

Lee, J. D., & See, K. A. (2018). Trust in Automation: Designing for Appropriate Reliance. *Human Factors*, *50*(1), 97–110.

Leung, D., Law, R., vanHoof, H., & Buhalis, D. (2013). Socialmedia in tourismandhospitality: A literaturereview. *Journal of Travel & Tourism Marketing*, *30*(1-2), 3–22. doi:10.1080/10548408.2013.750919

Liang, Y., Wang, X., & Hou, J. (2021). Understanding User Needs for Artificial Intelligence in Aviation: A User-Centered Design Perspective. *IEEE Access : Practical Innovations, Open Solutions*, *9*, 1838–1847.

Li, J., Bon, M. A., & Ye, B. H. (2019). Hotel employee's artificial intelligence and robotics awareness and its impact on turnover intention: The moderating roles of perceived organizational support and competitive psychological climate. *Tourism Management*, *73*, 172–181. doi:10.1016/j.tourman.2019.02.006

Linnes, C., & Metcalf, B. (2017). iGeneration and their acceptance of technology. [IJMIS]. *International Journal of Management & Information Systems*, *21*(2), 11–26.

Linz, M. (2012). Scenarios for the aviation industry: A Delphi-based analysis for 2025. *Journal of Air Transport Management*, *22*, 28–35. doi:10.1016/j.jairtraman.2012.01.006

Liu, Y., Li, Y., Wang, G., & Dong, Z. (2020). A Comprehensive Review of Predictive Maintenance: Issues and Challenges. *Applied Sciences (Basel, Switzerland)*, *10*(8), 2847.

Lu, C. T., Wetmore, M., & Przetak, R. (2006). Another approach to enhance airline safety: Using management safety tools. *Journal of Air Transportation (Reston, Va.)*, *11*(2), 113–139.

Lufthansa.com. (2022). *Lufthansa apps*. Luftansa. https://www.lufthansa.com/rs/en/lufthansaapps. 16.04.2022.

Luo, X. R., & Zhdanov, D. (2016). Special issue introduction: A comprehensive perspective on informationsystems security — technical advances and behavioral issues. *Decision Support Systems*, *92*, 1–2. doi:10.1016/j.dss.2016.10.003

Malighetti, P., Meoli, M., Paleari, S., & Redondi, R. (2011). Value determinants in the aviation industry. *Transportation Research Part E, Logistics and Transportation Review*, *47*(3), 359–370. doi:10.1016/j.tre.2010.11.007

Malik, N., Tripathi, S. N., Kar, A. K., & Gupta, S. (2021). Impact of artificial intelligence on employees working in industry 4.0 led organizations. *International Journal of Manpower*, *43*(2), 334–354. doi:10.1108/IJM-03-2021-0173

Mariani, M. M., DiFelice, M. & Mura, M. (2016). Facebook as a destination marketing tool: Evidence from Italian regional Destination Management Organizations. *Tourism Management, 54*, ss.321-343.

Marketingturkiye.com.tr. (2021, October 9). *Influencer Büyüsünü Güçlendirmek.* Marketing Turkiye. https://www.marketingturkiye.com.tr/haberler/influencer-buyusu/

Marketingturkiye.com.tr. (2023, March 26). *Sosyal Medyanın En Etkili Influencer'ları.* Marketing Turkiye.

Mayo, G., & Snider, D. (2016). *Bas and cyber security: A multiple discipline perspective.* In *2016 International Annual Conference of the American Society for Engineering Management*, Concord, NC, US.

McAndrew, I., Vishnevskaya, E., & Johnson, M. (2019). *Artificial intelligence in the aviation manufacturing process for complex assemblies and components.* In 2019 6th International Conference on Mechanical, Materials and Manufacturing, Boston, MA.

McDonnell, S., Nada, O., Abid, M. R., & Amjadian, E. (2021). CyberBERT: A deep dynamic-state session-based Recommender System for Cyber Threat Recognition. *2021 IEEE Aerospace Conference*. IEEE. 10.1109/AERO50100.2021.9438286

Merlo, T. R. (2024). *AI and data analytics applications in organizational management*. IGI-Global.

Miguéns, J., Baggio, R., & Costa, C. (2008). Social media and Tourism Destinations: TripAdvisor Case Study. *Advances in Tourism Research, Aveiro, Portugal*, (May), 26–28.

Mohamed, M. M., Gomaa, H. A. and El-Sherif, N. S. (2018). Evaluation of Current Smart Airport Technologies Implemented İn Cairo International Airport. *International Journal Of Heritage, Tourism And Hospitality, 12* (2/2), 130-140.

Monash University Hargrave-Andrew Library. (2006). *Remote Piloted Aerial Vehicles: An Anthology*. Monash University. https://www.ctie.monash.edu/hargrave/rpav_home.html#Beginnings

NATO UAS Classification Guide. September 2009 *Joint Capability Group on Unmanned Aerial Vehicles (JCGUAV) meeting*. NATO.

NEC. (2017). *Smart Airport Supports Your Smooth Travel.* NEC. Www.Nec.Com

Newcome, L. R. (2004). *Unmanned Aviation: A Brief History of Unmanned Aerial Vehicles.* American Institute of Aeronautics and Astronautics. doi:10.2514/4.868894

NIST. (2021). *Cyberspace*. NIST. https://csrc.nist.gov/glossary/term/cyberspace

Nothing artificial about this intelligence: Delta's industry-first machine learning platform minimizes customer inconvenience during tough operations. (2020). Delta. Retrieved October 4, 2023 from https://news.delta.com/nothing-artificial-about-intelligence-deltas-industry-first-machine-learning-platform-minimizes

Pabel, A., & Prideaux, B. (2016). Social Media Use İn Pre-Trip Planning By Tourists Visiting A Small Regional Leisure Destination. *Journal of Vacation Marketing*, *22*(4), 335–348. doi:10.1177/1356766715618998

Paek, H. J., Hove, T., Jung, Y., & Cole, R. T. (2013). Engagement across three social media platforms: An exploratory study of a cause-related PR campaign. *Public Relations Review*, *39*(5), 526–533. doi:10.1016/j.pubrev.2013.09.013

Parker, S. K., & Grote, G. (2022). Automation, algorithms, and beyond: Why work design matters more than ever in a digital world. *Applied Psychology*, *71*(4), 1171–1204. doi:10.1111/apps.12241

Patterson, R. E., Broy, N., & Leiden, K. L. (2022). User Empowerment in the Age of Artificial Intelligence: Designing for User Training and Interaction in Complex Systems. *Human Factors*, *54*(1), 5–24.

Peng, Y., Lu, T., Liu, J., Gao, Y., Guo, X., & Xie, F. (2013). Cyber-physical system risk assessment. In *2013 Ninth International Conference on Intelligent Information Hiding and Multimedia Signal Processing*. IEEE. 10.1109/IIH-MSP.2013.116

Petrenko, E. (2015). Designing effective pre-flight airport security information in electronic format: *Case study: Vilnius International Airport*. Theseus. https://www.theseus.fi/bitstream/handle/10024/101650/Petrenko_Ernestas2a.pdf?sequence=1

Polat, S., & Esen, F. (2021). Internet of Things For Travel Services. S. Polat & M. F. Esen (Eds.). Digital Technology Advancements In Knowledge Management. IGI Global.

Pop, R. A., Săplăcan, Z., Dabija, D. C., & Alt, M. A. (2022). The impact of social media influencers on travel decisions: The role of trust in consumer decision journey. *Current Issues in Tourism*, *25*(5), 823–843. doi:10.1080/13683500.2021.1895729

Porche, R. P. III. (2020). *Cyberwarfare: An Introduction to Information-Age Conflict*. Artech House.

Prensky, M. (2001). Digital natives, digital immigrants. *On the Horizon*, *9*(5), 1–6. doi:10.1108/10748120110424816

Punel, A., & Ermagun, A. (2018). Using Twitter network to detect market segments in the airline industry. *Journal of Air Transport Management*, *73*, 67–76. doi:10.1016/j.jairtraman.2018.08.004

Ranganathan, G. (2023). Airline CEO's AI system for driving personalization. *Journal of Revenue and Pricing Management*, *22*, 166–170. https://doi.org/10.1057/s41272-022-00402-w

Rawassizadeh, R., Price, B., & Petre, M. (2015). Wearables:Has The Age of Smartwatches Finally Arrived? *Communications of the ACM*, *58*(1), 45–47. doi:10.1145/2629633

Reis, M. F. (2022). *Evolution of the Aviation Industry: Impact of Digital Transformation of Learning Solutions*. HDL. http://hdl.handle.net/10400.6/13034

Revfine. (2021). Alexa for hospitality; voice control speakers for hotels. *RevFine*. https://www.revfine.com/alexa-for-hospitality/.

Rice, C., Ragbir, N. K., Rice, S., & Barcia, G. (2020). Willingness to pay for sustainable aviation depends on ticket price, greenhouse gas reductions and gender. *Technology in Society*, *60*, 101224. doi:10.1016/j.techsoc.2019.101224

Saballa, J. (2023, October 13). Israel military using new AI system in combat operations: report. The Defense Post. Retrieved from https://www.thedefensepost.com/2023/07/17/israel-military-ai-system/

Samatas, G. G., Moumgiakmas, S. S., & Papakostas, G. A. (2021, May). Predictive maintenance-bridging artificial intelligence and iot. In 2021 IEEE World AI IoT Congress (AIIoT); pp. 413-419. IEEE.

Sankaran, S., Müller, R., & Drouin, N. (2020). Creating a 'sustainability sublime' to enable megaprojects to meet the United Nations sustainable development goals. *Systems Research and Behavioral Science*, *37*(5), 813–826. doi:10.1002/sres.2744

Şanlı, S. C., Atçı, D., & Unur, K. (2019). Mersin'de Yiyecek İçecek İşletmelerinde Çalışan Personelin İşte Cep Telefonu Kullanımına İlişkin Algıları. *International Congress on "Dijital Transformation in Tourism I-DIGIT-19*, (pp. 26-42). IEEE.

Şanlıöz, K., Dilek, E. S., & Koçak, N. (2013). Değişen Dünya, Dönüşen Pazarlama: Türkiye Turizm Sektöründen Öncü Bir Mobil Uygulama Örneği. *Anatolia: Turizm Araştırmaları Dergisi*, *24*(2), 250–260.

Schmallegger, D., & Carson, D. (2008). Blogs in Tourism: Changing Approaches to Information Exchange. *Journal of Vacation Marketing*, *14*(2), 99–110. doi:10.1177/1356766707087519

Scott, B. I., & Trimarchi, A. (2017). The Digital Aviation Industry: A Balancing Act Between Cybersecurity and European Consumer Protection. *Air and Space Law*, *42*(4/5), 443–462.

Selenko, E., Bankins, S., Shoss, M., Warburton, J., & Restubog, S. (2022). Artificial intelligence and the future of work: A functional-identity perspective. *Current Directions in Psychological Science*, *31*(3), 272–279. doi:10.1177/09637214221091823

Singer, P. W. (2009). *Wired for War: The Robotics Revolution and Conflict in the Twenty-First Century*. Penguin Press.

Sloggett, D. (2015). *Drone Warfare: The Development of Unmanned Aerial Conflict*. Skyhorse Publishing.

Smith, A. M., Hancock, P. A., & Clegg, B. A. (2019). Task Load Management: Implications for Human-System Interaction. *Applied Ergonomics*, *74*, 234–242.

Smith, B. G., & Gallicano, T. D. (2015). Terms of engagement: Analyzing public engagement with organizations through social media. *Computers in Human Behavior*, *53*, 82–90. doi:10.1016/j.chb.2015.05.060

Smith, E., W., Kimbu, A. N., de Jong, A., & Cohen, S. (2023). Gendered Instagram representations in the aviation industry. *Journal of Sustainable Tourism, 31*(3), 639–663. doi:10.1080/09669582.2021.1932933

Sommer, P., & Brown, I. (2011). Reducing Systemic Cyber Security Risk. *Organisation for Economic Cooperation and Development Working Paper No. IFP/WKP/FGS, 2011*(3).

Song, H., & Liu, H. (2017). Predicting tourist demand using big data. In *Analytics in smart tourism design* (pp. 13–29). Springer. doi:10.1007/978-3-319-44263-1_2

Sridhar, S., Hahn, A., & Govindarasu, M. (2012). Cyber-physical system security for the electric power grid. *Proceedings of the IEEE, 100*(1), 210–224. doi:10.1109/JPROC.2011.2165269

Stanton, I., Munir, K., Ikram, A., & El-Bakry, M. (2023). Predictive maintenance analytics and implementation for aircraft: Challenges and opportunities. *Systems Engineering, 26*(2), 216–237.

SumithaK. (2023). "Exploring the Role of Artificial Intelligence in Improving Passenger Satisfaction in the Airline Industry: An Analysis of Customer Feedback and AI-Driven Solutions.". doi:10.5281/zenodo.7827975

Tian, X., He, W., Tang, C., Li, L., Xu, H., & Selover, D. (2020). A new approach of social media analytics to predict service quality: Evidence from the airline industry. *Journal of Enterprise Information Management, 33*(1), 51–70. doi:10.1108/JEIM-03-2019-0086

Tikhonov, A. I., Sazonov, A. A., & Novikov, S. V. (2019). Digital aviation industry in Russia. *Russian Engineering Research, 39*(4), 349–353. doi:10.3103/S1068798X19040178

Turel, O., Matt, C., Trenz, M., Cheung, C. M., D'Arcy, J., Qahri-Saremi, H., & Tarafdar, M. (2019). Panel report: The dark side of the digitization of the individual. *Internet Research, 29*(2), 274–288. doi:10.1108/INTR-04-2019-541

Türsab & TBV. (2020). Turizm Sektörü Dijitalleşme Yol Haritası. Tursab. https://www.tursab.org.tr/apps/Files/Content/ad5f3ddb-5a11-410f-9e3cfe8b2dc4df8b.pdf,.

Türsab (2020). *Türkiye ve Dünya Turizmi Değerlendirmesi*. Tursab.

Urry, J. (1999). *Mekânları Tüketmek*. Mart Yayıncılık.

Van Bossuyt, D. L., & O'Halloran, B. M. (2019). A method to choose between automation and human operators for recovery actions during a cyber attack. *Procedia Computer Science, 153*, 352–360. doi:10.1016/j.procs.2019.05.089

Van Der Schaft, A. H., Lub, X. D., Van Der Heijden, B., & Solinger, O. N. (2022). How employees experience digital transformation: A dynamic and multi-layered sensemaking perspective. *Journal of Hospitality & Tourism Research (Washington, D.C.)*, 1–18. doi:10.1177/10963480221123098

Wagner, S. M., & Walton, R. O. (2016). Additive manufacturing's impact and future in the aviation industry. *Production Planning and Control, 27*(13), 1124–1130. doi:10.1080/09537287.2016.1199824

Wang, D. & ve Xiang, Z. (2012). The New Landscape of Travel: A Comprehensive Analysis of Smartphone Apps. In M. Fuchs, F. Ricci ve L. Cantoni (Eds.), Information and Communication Technologies in Tourism. Wien: Springer.

Wang, S., Ding, L., Sui, H., & Gu, Z. (2021). Cybersecurity risk assessment method of ICS based on attack-defense tree model. *Journal of Intelligent & Fuzzy Systems*, *40*(6), 10475–10488. doi:10.3233/JIFS-201126

Wang, X., & Liu, Z. (2019). Online engagement in social media: A cross-cultural comparison. *Computers in Human Behavior*, *97*, 137–150. doi:10.1016/j.chb.2019.03.014

Want, R. (2006). An Introduction To RFID Technology. *IEEE Pervasive Computing*, *5*(1), 25–33. doi:10.1109/MPRV.2006.2

Wardani, P. E., Marroy, D. S. H., Octora, Y., & Setiawan, E. B. (2019). Impact of Social Media Influencer Towards Brand Attitude and Purchase Intention in Airline Industry. *Advances in Transportation and Logistics Research*, *2*, 496–503.

Warkentin, M., Johnston, A. C., Shropshire, J., & Barnett, W. D. (2016). Continuance of protective security behavior: A longitudinal study. *Decision Support Systems*, *92*, 25–35. doi:10.1016/j.dss.2016.09.013

Watts, A. C., Ambrosia, V. G., & Hinkley, E. A. (2012). Unmanned Aircraft Systems in Remote Sensing and Scientific Research: Classification and Considerations of Use. *Remote Sensing (Basel)*, *4*(6), 1671–1692. doi:10.3390/rs4061671

Weckman, G. R., Shell, R. L., & Marvel, J. H. (2001). Modeling the reliability of repairable systems in the aviation industry. *Computers & Industrial Engineering*, *40*(1-2), 51–63. doi:10.1016/S0360-8352(00)00063-2

Weibel, R. E., & Hansman, R. J. (2004). Safety considerations for operation of different classes of UAVs in the NAS. *Proceedings of the 4th Aviation Technology, Integration and Operations Forum, AIAA 3rd Unmanned Unlimited Technical Conference, Workshop and Exhibit*. IEEE.

Westwood, S., Pritchard, A., & Morgan, N. (2000). Gender-Blind Marketing: Businesswomen's Perceptions of Airline Services. *Tourism Management*, *21*(4), 353–362. doi:10.1016/S0261-5177(99)00069-2

Whitmore, B. A. (2016). *Evolution of Unmanned Aerial Warfare: A Historical Look at Remote Airpower—A Case Study in Innovation*. [Thesis, U.S. Army Command and General Staff College, Fort Leavenworth, KS].

Wibisono, A. I., & Ruldeviyani, Y. (2021, April). Detecting Social Media Influencers of Airline Services through Social Network Analysis on Twitter: A Case Study of the Indonesian Airline Industry. In *2021 3rd East Indonesia Conference on Computer and Information Technology (EIConCIT)* (pp. 314-319). IEEE. 10.1109/EIConCIT50028.2021.9431876

Wilson, J., Sachdev, S., & Alter, A. (2016). How Companies Are Using Machine Learning to Get Faster and More Efficient. Harvard Business Review. Retrieved on October 4, 2023 from: https://hbr.org/2016/05/how-companies-are-using-machine-learning-to-get-faster-and-more-efficient

Wilson, A. D., & Russell, D. M. (2019). Human Performance with Control Room Automation: A Literature Review and Meta-Analysis. *Applied Ergonomics*, *76*, 192–204.

Wittmer, A. (2011). Acceptance Of Self-Service Check-İn At Zurich Airport. *Research in Transportation Business & Management*, *1*(1), 136–143. doi:10.1016/j.rtbm.2011.06.001

Wong, K., Rudd, C., Pickering, S., & Liu, X. (2017). Composites recycling solutions for the aviation industry. *Science China. Technological Sciences*, *60*(9), 1291–1300. doi:10.1007/s11431-016-9028-7

Woods, D., & Dekker, S. (2000). Anticipating the effects of technological change: A new era of dynamics for human factors. *Theoretical Issues in Ergonomics Science*, *1*(3), 272–282. doi:10.1080/14639220110037452

Wu, W. (2020). *Analysis of Digital Tourism, Virtual Tourism and Wisdom Tourism.* Springer. https://link.springer.com/conference/csia

Xiong, M., & Wang, H. (2022). Digital twin applications in aviation industry: A review. *International Journal of Advanced Manufacturing Technology*, *121*(9-10), 5677–5692. doi:10.1007/s00170-022-09717-9

Yalçın Kayıkçı, M., & Kutluk Bozkurt, A. (2018). Dijital Çağda Z Ve Alpha Kuşağı, Yapay Zeka Uygulamaları Ve Turizme Yansımaları. *Sosyal Bilimler Metinleri*, *01*, 54–64.

Zaloga, S. (2008). *Unmanned Aerial Vehicles: Robotic Air Warfare, 1917-2007.* Osprey Publishing.

Катерна, О. К., & Молчанова, К. М. (2020). Digital Transformation of Ransformation of Aviation Industry in Ukraine. *ЕКОНОМІКА І УПРАВЛІННЯ*, 53-63.

# Related References

To continue our tradition of advancing academic research, we have compiled a list of recommended IGI Global readings. These references will provide additional information and guidance to further enrich your knowledge and assist you with your own research and future publications.

Abbasnejad, B., Moeinzadeh, S., Ahankoob, A., & Wong, P. S. (2021). The Role of Collaboration in the Implementation of BIM-Enabled Projects. In J. Underwood & M. Shelbourn (Eds.), *Handbook of Research on Driving Transformational Change in the Digital Built Environment* (pp. 27–62). IGI Global. https://doi.org/10.4018/978-1-7998-6600-8.ch002

Abdulrahman, K. O., Mahamood, R. M., & Akinlabi, E. T. (2022). Additive Manufacturing (AM): Processing Technique for Lightweight Alloys and Composite Material. In K. Kumar, B. Babu, & J. Davim (Ed.), *Handbook of Research on Advancements in the Processing, Characterization, and Application of Lightweight Materials* (pp. 27-48). IGI Global. https://doi.org/10.4018/978-1-7998-7864-3.ch002

Agrawal, R., Sharma, P., & Saxena, A. (2021). A Diamond Cut Leather Substrate Antenna for BAN (Body Area Network) Application. In V. Singh, V. Dubey, A. Saxena, R. Tiwari, & H. Sharma (Eds.), *Emerging Materials and Advanced Designs for Wearable Antennas* (pp. 54–59). IGI Global. https://doi.org/10.4018/978-1-7998-7611-3.ch004

Ahmad, F., Al-Ammar, E. A., & Alsaidan, I. (2022). Battery Swapping Station: A Potential Solution to Address the Limitations of EV Charging Infrastructure. In M. Alam, R. Pillai, & N. Murugesan (Eds.), *Developing Charging Infrastructure and Technologies for Electric Vehicles* (pp. 195–207). IGI Global. doi:10.4018/978-1-7998-6858-3.ch010

Aikhuele, D. (2018). A Study of Product Development Engineering and Design Reliability Concerns. *International Journal of Applied Industrial Engineering*, *5*(1), 79–89. doi:10.4018/IJAIE.2018010105

Al-Khatri, H., & Al-Atrash, F. (2021). Occupants' Habits and Natural Ventilation in a Hot Arid Climate. In R. González-Lezcano (Ed.), *Advancements in Sustainable Architecture and Energy Efficiency* (pp. 146–168). IGI Global. https://doi.org/10.4018/978-1-7998-7023-4.ch007

Al-Shebeeb, O. A., Rangaswamy, S., Gopalakrishan, B., & Devaru, D. G. (2017). Evaluation and Indexing of Process Plans Based on Electrical Demand and Energy Consumption. *International Journal of Manufacturing, Materials, and Mechanical Engineering*, *7*(3), 1–19. doi:10.4018/IJMMME.2017070101

Amuda, M. O., Lawal, T. F., & Akinlabi, E. T. (2017). Research Progress on Rheological Behavior of AA7075 Aluminum Alloy During Hot Deformation. *International Journal of Materials Forming and Machining Processes*, *4*(1), 53–96. doi:10.4018/IJMFMP.2017010104

Amuda, M. O., Lawal, T. F., & Mridha, S. (2021). Microstructure and Mechanical Properties of Silicon Carbide-Treated Ferritic Stainless Steel Welds. In L. Burstein (Ed.), *Handbook of Research on Advancements in Manufacturing, Materials, and Mechanical Engineering* (pp. 395–411). IGI Global. https://doi.org/10.4018/978-1-7998-4939-1.ch019

Anikeev, V., Gasem, K. A., & Fan, M. (2021). Application of Supercritical Technologies in Clean Energy Production: A Review. In L. Chen (Ed.), *Handbook of Research on Advancements in Supercritical Fluids Applications for Sustainable Energy Systems* (pp. 792–821). IGI Global. https://doi.org/10.4018/978-1-7998-5796-9.ch022

Arafat, M. Y., Saleem, I., & Devi, T. P. (2022). Drivers of EV Charging Infrastructure Entrepreneurship in India. In M. Alam, R. Pillai, & N. Murugesan (Eds.), *Developing Charging Infrastructure and Technologies for Electric Vehicles* (pp. 208–219). IGI Global. https://doi.org/10.4018/978-1-7998-6858-3.ch011

Araujo, A., & Manninen, H. (2022). Contribution of Project-Based Learning on Social Skills Development: An Industrial Engineer Perspective. In A. Alves & N. van Hattum-Janssen (Eds.), *Training Engineering Students for Modern Technological Advancement* (pp. 119–145). IGI Global. https://doi.org/10.4018/978-1-7998-8816-1.ch006

Armutlu, H. (2018). Intelligent Biomedical Engineering Operations by Cloud Computing Technologies. In U. Kose, G. Guraksin, & O. Deperlioglu (Eds.), *Nature-Inspired Intelligent Techniques for Solving Biomedical Engineering Problems* (pp. 297–317). Hershey, PA: IGI Global. doi:10.4018/978-1-5225-4769-3.ch015

Atik, M., Sadek, M., & Shahrour, I. (2017). Single-Run Adaptive Pushover Procedure for Shear Wall Structures. In V. Plevris, G. Kremmyda, & Y. Fahjan (Eds.), *Performance-Based Seismic Design of Concrete Structures and Infrastructures* (pp. 59–83). Hershey, PA: IGI Global. doi:10.4018/978-1-5225-2089-4.ch003

Attia, H. (2021). Smart Power Microgrid Impact on Sustainable Building. In R. González-Lezcano (Ed.), *Advancements in Sustainable Architecture and Energy Efficiency* (pp. 169–194). IGI Global. https://doi.org/10.4018/978-1-7998-7023-4.ch008

Aydin, A., Akyol, E., Gungor, M., Kaya, A., & Tasdelen, S. (2018). Geophysical Surveys in Engineering Geology Investigations With Field Examples. In N. Ceryan (Ed.), *Handbook of Research on Trends and Digital Advances in Engineering Geology* (pp. 257–280). Hershey, PA: IGI Global. doi:10.4018/978-1-5225-2709-1.ch007

Ayoobkhan, M. U. D., Y., A., J., Easwaran, B., & R., T. (2021). Smart Connected Digital Products and IoT Platform With the Digital Twin. In P. Vasant, G. Weber, & W. Punurai (Ed.), Research Advancements in Smart Technology, Optimization, and Renewable Energy (pp. 330-350). IGI Global. https://doi.org/ doi:10.4018/978-1-7998-3970-5.ch016

Baeza Moyano, D., & González Lezcano, R. A. (2021). The Importance of Light in Our Lives: Towards New Lighting in Schools. In R. González-Lezcano (Ed.), *Advancements in Sustainable Architecture and Energy Efficiency* (pp. 239–256). IGI Global. https://doi.org/10.4018/978-1-7998-7023-4.ch011

Bagdadee, A. H. (2021). A Brief Assessment of the Energy Sector of Bangladesh. *International Journal of Energy Optimization and Engineering, 10*(1), 36–55. doi:10.4018/IJEOE.2021010103

Baklezos, A. T., & Hadjigeorgiou, N. G. (2021). Magnetic Sensors for Space Applications and Magnetic Cleanliness Considerations. In C. Nikolopoulos (Ed.), *Recent Trends on Electromagnetic Environmental Effects for Aeronautics and Space Applications* (pp. 147–185). IGI Global. https://doi.org/10.4018/978-1-7998-4879-0.ch006

Bas, T. G. (2017). Nutraceutical Industry with the Collaboration of Biotechnology and Nutrigenomics Engineering: The Significance of Intellectual Property in the Entrepreneurship and Scientific Research Ecosystems. In T. Bas & J. Zhao (Eds.), *Comparative Approaches to Biotechnology Development and Use in Developed and Emerging Nations* (pp. 1–17). Hershey, PA: IGI Global. doi:10.4018/978-1-5225-1040-6.ch001

Bazeer Ahamed, B., & Periakaruppan, S. (2021). Taxonomy of Influence Maximization Techniques in Unknown Social Networks. In P. Vasant, G. Weber, & W. Punurai (Eds.), *Research Advancements in Smart Technology, Optimization, and Renewable Energy* (pp. 351-363). IGI Global. https://doi.org/10.4018/978-1-7998-3970-5.ch017

Beale, R., & André, J. (2017). *Design Solutions and Innovations in Temporary Structures*. Hershey, PA: IGI Global. doi:10.4018/978-1-5225-2199-0

Behnam, B. (2017). Simulating Post-Earthquake Fire Loading in Conventional RC Structures. In P. Samui, S. Chakraborty, & D. Kim (Eds.), *Modeling and Simulation Techniques in Structural Engineering* (pp. 425–444). Hershey, PA: IGI Global. doi:10.4018/978-1-5225-0588-4.ch015

Ben Hamida, I., Salah, S. B., Msahli, F., & Mimouni, M. F. (2018). Distribution Network Reconfiguration Using SPEA2 for Power Loss Minimization and Reliability Improvement. *International Journal of Energy Optimization and Engineering*, *7*(1), 50–65. doi:10.4018/IJEOE.2018010103

Bentarzi, H. (2021). Fault Tree-Based Root Cause Analysis Used to Study Mal-Operation of a Protective Relay in a Smart Grid. In A. Recioui & H. Bentarzi (Eds.), *Optimizing and Measuring Smart Grid Operation and Control* (pp. 289–308). IGI Global. https://doi.org/10.4018/978-1-7998-4027-5.ch012

Beysens, D. A., Garrabos, Y., & Zappoli, B. (2021). Thermal Effects in Near-Critical Fluids: Piston Effect and Related Phenomena. In L. Chen (Ed.), *Handbook of Research on Advancements in Supercritical Fluids Applications for Sustainable Energy Systems* (pp. 1–31). IGI Global. https://doi.org/10.4018/978-1-7998-5796-9.ch001

Bhaskar, S. V., & Kudal, H. N. (2017). Effect of TiCN and AlCrN Coating on Tribological Behaviour of Plasma-nitrided AISI 4140 Steel. *International Journal of Surface Engineering and Interdisciplinary Materials Science*, *5*(2), 1–17. doi:10.4018/IJSEIMS.2017070101

Bhuyan, D. (2018). Designing of a Twin Tube Shock Absorber: A Study in Reverse Engineering. In K. Kumar & J. Davim (Eds.), *Design and Optimization of Mechanical Engineering Products* (pp. 83–104). Hershey, PA: IGI Global. doi:10.4018/978-1-5225-3401-3.ch005

Blumberg, G. (2021). Blockchains for Use in Construction and Engineering Projects. In J. Underwood & M. Shelbourn (Eds.), *Handbook of Research on Driving Transformational Change in the Digital Built Environment* (pp. 179–208). IGI Global. https://doi.org/10.4018/978-1-7998-6600-8.ch008

Bolboaca, A. M. (2021). Considerations Regarding the Use of Fuel Cells in Combined Heat and Power for Stationary Applications. In G. Badea, R. Felseghi, & I. Aşchilean (Eds.), *Hydrogen Fuel Cell Technology for Stationary Applications* (pp. 239–275). IGI Global. https://doi.org/10.4018/978-1-7998-4945-2.ch010

Burstein, L. (2021). Simulation Tool for Cable Design. In L. Burstein (Ed.), *Handbook of Research on Advancements in Manufacturing, Materials, and Mechanical Engineering* (pp. 54–74). IGI Global. https://doi.org/10.4018/978-1-7998-4939-1.ch003

Calderon, F. A., Giolo, E. G., Frau, C. D., Rengel, M. G., Rodriguez, H., Tornello, M., ... Gallucci, R. (2018). Seismic Microzonation and Site Effects Detection Through Microtremors Measures: A Review. In N. Ceryan (Ed.), *Handbook of Research on Trends and Digital Advances in Engineering Geology* (pp. 326–349). Hershey, PA: IGI Global. doi:10.4018/978-1-5225-2709-1.ch009

Ceryan, N., & Can, N. K. (2018). Prediction of The Uniaxial Compressive Strength of Rocks Materials. In N. Ceryan (Ed.), *Handbook of Research on Trends and Digital Advances in Engineering Geology* (pp. 31–96). Hershey, PA: IGI Global. doi:10.4018/978-1-5225-2709-1.ch002

Ceryan, S. (2018). Weathering Indices Used in Evaluation of the Weathering State of Rock Material. In N. Ceryan (Ed.), *Handbook of Research on Trends and Digital Advances in Engineering Geology* (pp. 132–186). Hershey, PA: IGI Global. doi:10.4018/978-1-5225-2709-1.ch004

Chen, H., Padilla, R. V., & Besarati, S. (2017). Supercritical Fluids and Their Applications in Power Generation. In L. Chen & Y. Iwamoto (Eds.), *Advanced Applications of Supercritical Fluids in Energy Systems* (pp. 369–402). Hershey, PA: IGI Global. doi:10.4018/978-1-5225-2047-4.ch012

Chen, H., Padilla, R. V., & Besarati, S. (2021). Supercritical Fluids and Their Applications in Power Generation. In L. Chen (Ed.), *Handbook of Research on Advancements in Supercritical Fluids Applications for Sustainable Energy Systems* (pp. 566–599). IGI Global. https://doi.org/10.4018/978-1-7998-5796-9.ch016

Chen, L. (2017). Principles, Experiments, and Numerical Studies of Supercritical Fluid Natural Circulation System. In L. Chen & Y. Iwamoto (Eds.), *Advanced Applications of Supercritical Fluids in Energy Systems* (pp. 136–187). Hershey, PA: IGI Global. doi:10.4018/978-1-5225-2047-4.ch005

Chen, L. (2021). Principles, Experiments, and Numerical Studies of Supercritical Fluid Natural Circulation System. In L. Chen (Ed.), *Handbook of Research on Advancements in Supercritical Fluids Applications for Sustainable Energy Systems* (pp. 219–269). IGI Global. https://doi.org/10.4018/978-1-7998-5796-9.ch007

Chiba, Y., Marif, Y., Henini, N., & Tlemcani, A. (2021). Modeling of Magnetic Refrigeration Device by Using Artificial Neural Networks Approach. *International Journal of Energy Optimization and Engineering, 10*(4), 68–76. https://doi.org/10.4018/IJEOE.2021100105

Clementi, F., Di Sciascio, G., Di Sciascio, S., & Lenci, S. (2017). Influence of the Shear-Bending Interaction on the Global Capacity of Reinforced Concrete Frames: A Brief Overview of the New Perspectives. In V. Plevris, G. Kremmyda, & Y. Fahjan (Eds.), *Performance-Based Seismic Design of Concrete Structures and Infrastructures* (pp. 84–111). Hershey, PA: IGI Global. doi:10.4018/978-1-5225-2089-4.ch004

Codinhoto, R., Fialho, B. C., Pinti, L., & Fabricio, M. M. (2021). BIM and IoT for Facilities Management: Understanding Key Maintenance Issues. In J. Underwood & M. Shelbourn (Eds.), *Handbook of Research on Driving Transformational Change in the Digital Built Environment* (pp. 209–231). IGI Global. doi:10.4018/978-1-7998-6600-8.ch009

Cortés-Polo, D., Calle-Cancho, J., Carmona-Murillo, J., & González-Sánchez, J. (2017). Future Trends in Mobile-Fixed Integration for Next Generation Networks: Classification and Analysis. *International Journal of Vehicular Telematics and Infotainment Systems, 1*(1), 33–53. doi:10.4018/IJVTIS.2017010103

Costa, H. G., Sheremetieff, F. H., & Araújo, E. A. (2022). Influence of Game-Based Methods in Developing Engineering Competences. In A. Alves & N. van Hattum-Janssen (Eds.), *Training Engineering Students for Modern Technological Advancement* (pp. 69–88). IGI Global. https://doi.org/10.4018/978-1-7998-8816-1.ch004

Cui, X., Zeng, S., Li, Z., Zheng, Q., Yu, X., & Han, B. (2018). Advanced Composites for Civil Engineering Infrastructures. In K. Kumar & J. Davim (Eds.), *Composites and Advanced Materials for Industrial Applications* (pp. 212–248). Hershey, PA: IGI Global. doi:10.4018/978-1-5225-5216-1.ch010

Dalgıç, S., & Kuşku, İ. (2018). Geological and Geotechnical Investigations in Tunneling. In N. Ceryan (Ed.), *Handbook of Research on Trends and Digital Advances in Engineering Geology* (pp. 482–529). Hershey, PA: IGI Global. doi:10.4018/978-1-5225-2709-1.ch014

Dang, C., & Hihara, E. (2021). Study on Cooling Heat Transfer of Supercritical Carbon Dioxide Applied to Transcritical Carbon Dioxide Heat Pump. In L. Chen (Ed.), *Handbook of Research on Advancements in Supercritical Fluids Applications for Sustainable Energy Systems* (pp. 451–493). IGI Global. https://doi.org/10.4018/978-1-7998-5796-9.ch013

Daus, Y., Kharchenko, V., & Yudaev, I. (2021). Research of Solar Energy Potential of Photovoltaic Installations on Enclosing Structures of Buildings. *International Journal of Energy Optimization and Engineering, 10*(4), 18–34. https://doi.org/10.4018/IJEOE.2021100102

Daus, Y., Kharchenko, V., & Yudaev, I. (2021). Optimizing Layout of Distributed Generation Sources of Power Supply System of Agricultural Object. *International Journal of Energy Optimization and Engineering, 10*(3), 70–84. https://doi.org/10.4018/IJEOE.2021070104

de la Varga, D., Soto, M., Arias, C. A., van Oirschot, D., Kilian, R., Pascual, A., & Álvarez, J. A. (2017). Constructed Wetlands for Industrial Wastewater Treatment and Removal of Nutrients. In Á. Val del Río, J. Campos Gómez, & A. Mosquera Corral (Eds.), *Technologies for the Treatment and Recovery of Nutrients from Industrial Wastewater* (pp. 202–230). Hershey, PA: IGI Global. doi:10.4018/978-1-5225-1037-6.ch008

Deb, S., Ammar, E. A., AlRajhi, H., Alsaidan, I., & Shariff, S. M. (2022). V2G Pilot Projects: Review and Lessons Learnt. In M. Alam, R. Pillai, & N. Murugesan (Eds.), *Developing Charging Infrastructure and Technologies for Electric Vehicles* (pp. 252–267). IGI Global. https://doi.org/10.4018/978-1-7998-6858-3.ch014

Dekhandji, F. Z., & Rais, M. C. (2021). A Comparative Study of Power Quality Monitoring Using Various Techniques. In A. Recioui & H. Bentarzi (Eds.), *Optimizing and Measuring Smart Grid Operation and Control* (pp. 259–288). IGI Global. https://doi.org/10.4018/978-1-7998-4027-5.ch011

Deperlioglu, O. (2018). Intelligent Techniques Inspired by Nature and Used in Biomedical Engineering. In U. Kose, G. Guraksin, & O. Deperlioglu (Eds.), *Nature-Inspired Intelligent Techniques for Solving Biomedical Engineering Problems* (pp. 51–77). Hershey, PA: IGI Global. doi:10.4018/978-1-5225-4769-3.ch003

Dhurpate, P. R., & Tang, H. (2021). Quantitative Analysis of the Impact of Inter-Line Conveyor Capacity for Throughput of Manufacturing Systems. *International Journal of Manufacturing, Materials, and Mechanical Engineering*, *11*(1), 1–17. https://doi.org/10.4018/IJMMME.2021010101

Dinkar, S., & Deep, K. (2021). A Survey of Recent Variants and Applications of Antlion Optimizer. *International Journal of Energy Optimization and Engineering*, *10*(2), 48–73. doi:10.4018/IJEOE.2021040103

Dixit, A. (2018). Application of Silica-Gel-Reinforced Aluminium Composite on the Piston of Internal Combustion Engine: Comparative Study of Silica-Gel-Reinforced Aluminium Composite Piston With Aluminium Alloy Piston. In K. Kumar & J. Davim (Eds.), *Composites and Advanced Materials for Industrial Applications* (pp. 63–98). Hershey, PA: IGI Global. doi:10.4018/978-1-5225-5216-1.ch004

Drabecki, M. P., & Kułak, K. B. (2021). Global Pandemics on European Electrical Energy Markets: Lessons Learned From the COVID-19 Outbreak. *International Journal of Energy Optimization and Engineering*, *10*(3), 24–46. https://doi.org/10.4018/IJEOE.2021070102

Dutta, M. M. (2021). Nanomaterials for Food and Agriculture. In M. Bhat, I. Wani, & S. Ashraf (Eds.), *Applications of Nanomaterials in Agriculture, Food Science, and Medicine* (pp. 75–97). IGI Global. doi:10.4018/978-1-7998-5563-7.ch004

Dutta, M. M., & Goswami, M. (2021). Coating Materials: Nano-Materials. In S. Roy & G. Bose (Eds.), *Advanced Surface Coating Techniques for Modern Industrial Applications* (pp. 1–30). IGI Global. doi:10.4018/978-1-7998-4870-7.ch001

Elsayed, A. M., Dakkama, H. J., Mahmoud, S., Al-Dadah, R., & Kaialy, W. (2017). Sustainable Cooling Research Using Activated Carbon Adsorbents and Their Environmental Impact. In T. Kobayashi (Ed.), *Applied Environmental Materials Science for Sustainability* (pp. 186–221). Hershey, PA: IGI Global. doi:10.4018/978-1-5225-1971-3.ch009

Ercanoglu, M., & Sonmez, H. (2018). General Trends and New Perspectives on Landslide Mapping and Assessment Methods. In N. Ceryan (Ed.), *Handbook of Research on Trends and Digital Advances in Engineering Geology* (pp. 350–379). Hershey, PA: IGI Global. doi:10.4018/978-1-5225-2709-1.ch010

Faroz, S. A., Pujari, N. N., Rastogi, R., & Ghosh, S. (2017). Risk Analysis of Structural Engineering Systems Using Bayesian Inference. In P. Samui, S. Chakraborty, & D. Kim (Eds.), *Modeling and Simulation Techniques in Structural Engineering* (pp. 390–424). Hershey, PA: IGI Global. doi:10.4018/978-1-5225-0588-4.ch014

Fekik, A., Hamida, M. L., Denoun, H., Azar, A. T., Kamal, N. A., Vaidyanathan, S., Bousbaine, A., & Benamrouche, N. (2022). Multilevel Inverter for Hybrid Fuel Cell/PV Energy Conversion System. In A. Fekik & N. Benamrouche (Eds.), *Modeling and Control of Static Converters for Hybrid Storage Systems* (pp. 233–270). IGI Global. https://doi.org/10.4018/978-1-7998-7447-8.ch009

Fekik, A., Hamida, M. L., Houassine, H., Azar, A. T., Kamal, N. A., Denoun, H., Vaidyanathan, S., & Sambas, A. (2022). Power Quality Improvement for Grid-Connected Photovoltaic Panels Using Direct Power Control. In A. Fekik & N. Benamrouche (Eds.), *Modeling and Control of Static Converters for Hybrid Storage Systems* (pp. 107–142). IGI Global. https://doi.org/10.4018/978-1-7998-7447-8.ch005

Fernando, P. R., Hamigah, T., Disne, S., Wickramasingha, G. G., & Sutharshan, A. (2018). The Evaluation of Engineering Properties of Low Cost Concrete Blocks by Partial Doping of Sand with Sawdust: Low Cost Sawdust Concrete Block. *International Journal of Strategic Engineering*, *1*(2), 26–42. doi:10.4018/IJoSE.2018070103

Ferro, G., Minciardi, R., Parodi, L., & Robba, M. (2022). Optimal Charging Management of Microgrid-Integrated Electric Vehicles. In M. Alam, R. Pillai, & N. Murugesan (Eds.), *Developing Charging Infrastructure and Technologies for Electric Vehicles* (pp. 133–155). IGI Global. https://doi.org/10.4018/978-1-7998-6858-3.ch007

Flumerfelt, S., & Green, C. (2022). Graduate Lean Leadership Education: A Case Study of a Program. In A. Alves & N. van Hattum-Janssen (Eds.), *Training Engineering Students for Modern Technological Advancement* (pp. 202–224). IGI Global. https://doi.org/10.4018/978-1-7998-8816-1.ch010

Galli, B. J. (2021). Implications of Economic Decision Making to the Project Manager. *International Journal of Strategic Engineering*, *4*(1), 19–32. https://doi.org/10.4018/IJoSE.2021010102

Gento, A. M., Pimentel, C., & Pascual, J. A. (2022). Teaching Circular Economy and Lean Management in a Learning Factory. In A. Alves & N. van Hattum-Janssen (Eds.), *Training Engineering Students for Modern Technological Advancement* (pp. 183–201). IGI Global. https://doi.org/10.4018/978-1-7998-8816-1.ch009

Ghosh, S., Mitra, S., Ghosh, S., & Chakraborty, S. (2017). Seismic Reliability Analysis in the Framework of Metamodelling Based Monte Carlo Simulation. In P. Samui, S. Chakraborty, & D. Kim (Eds.), *Modeling and Simulation Techniques in Structural Engineering* (pp. 192–208). Hershey, PA: IGI Global. doi:10.4018/978-1-5225-0588-4.ch006

Gil, M., & Otero, B. (2017). Learning Engineering Skills through Creativity and Collaboration: A Game-Based Proposal. In R. Alexandre Peixoto de Queirós & M. Pinto (Eds.), *Gamification-Based E-Learning Strategies for Computer Programming Education* (pp. 14–29). Hershey, PA: IGI Global. doi:10.4018/978-1-5225-1034-5.ch002

Gill, J., Ayre, M., & Mills, J. (2017). Revisioning the Engineering Profession: How to Make It Happen! In M. Gray & K. Thomas (Eds.), *Strategies for Increasing Diversity in Engineering Majors and Careers* (pp. 156–175). Hershey, PA: IGI Global. doi:10.4018/978-1-5225-2212-6.ch008

Godzhaev, Z., Senkevich, S., Kuzmin, V., & Melikov, I. (2021). Use of the Neural Network Controller of Sprung Mass to Reduce Vibrations From Road Irregularities. In P. Vasant, G. Weber, & W. Punurai (Ed.), *Research Advancements in Smart Technology, Optimization, and Renewable Energy* (pp. 69-87). IGI Global. https://doi.org/10.4018/978-1-7998-3970-5.ch005

Gomes de Gusmão, C. M. (2022). Digital Competencies and Transformation in Higher Education: Upskilling With Extension Actions. In A. Alves & N. van Hattum-Janssen (Eds.), *Training Engineering Students for Modern Technological Advancement* (pp. 313–328). IGI Global. https://doi.org/10.4018/978-1-7998-8816-1.ch015A

Goyal, N., Ram, M., & Kumar, P. (2017). Welding Process under Fault Coverage Approach for Reliability and MTTF. In M. Ram & J. Davim (Eds.), *Mathematical Concepts and Applications in Mechanical Engineering and Mechatronics* (pp. 222–245). Hershey, PA: IGI Global. doi:10.4018/978-1-5225-1639-2.ch011

Gray, M., & Lundy, C. (2017). Engineering Study Abroad: High Impact Strategy for Increasing Access. In M. Gray & K. Thomas (Eds.), *Strategies for Increasing Diversity in Engineering Majors and Careers* (pp. 42–59). Hershey, PA: IGI Global. doi:10.4018/978-1-5225-2212-6.ch003

Güler, O., & Varol, T. (2021). Fabrication of Functionally Graded Metal and Ceramic Powders Synthesized by Electroless Deposition. In S. Roy & G. Bose (Eds.), *Advanced Surface Coating Techniques for Modern Industrial Applications* (pp. 150–187). IGI Global. https://doi.org/10.4018/978-1-7998-4870-7.ch007

Guraksin, G. E. (2018). Internet of Things and Nature-Inspired Intelligent Techniques for the Future of Biomedical Engineering. In U. Kose, G. Guraksin, & O. Deperlioglu (Eds.), *Nature-Inspired Intelligent Techniques for Solving Biomedical Engineering Problems* (pp. 263–282). Hershey, PA: IGI Global. doi:10.4018/978-1-5225-4769-3.ch013

Hamida, M. L., Fekik, A., Denoun, H., Ardjal, A., & Bokhtache, A. A. (2022). Flying Capacitor Inverter Integration in a Renewable Energy System. In A. Fekik & N. Benamrouche (Eds.), *Modeling and Control of Static Converters for Hybrid Storage Systems* (pp. 287–306). IGI Global. https://doi.org/10.4018/978-1-7998-7447-8.ch011

Hasegawa, N., & Takahashi, Y. (2021). Control of Soap Bubble Ejection Robot Using Facial Expressions. *International Journal of Manufacturing, Materials, and Mechanical Engineering, 11*(2), 1–16. https://doi.org/10.4018/IJMMME.2021040101

Hejazi, T., & Akbari, L. (2017). A Multiresponse Optimization Model for Statistical Design of Processes with Discrete Variables. In M. Ram & J. Davim (Eds.), *Mathematical Concepts and Applications in Mechanical Engineering and Mechatronics* (pp. 17–37). Hershey, PA: IGI Global. doi:10.4018/978-1-5225-1639-2.ch002

Hejazi, T., & Hejazi, A. (2017). Monte Carlo Simulation for Reliability-Based Design of Automotive Complex Subsystems. In M. Ram & J. Davim (Eds.), *Mathematical Concepts and Applications in Mechanical Engineering and Mechatronics* (pp. 177–200). Hershey, PA: IGI Global. doi:10.4018/978-1-5225-1639-2.ch009

Hejazi, T., & Poursabbagh, H. (2017). Reliability Analysis of Engineering Systems: An Accelerated Life Testing for Boiler Tubes. In M. Ram & J. Davim (Eds.), *Mathematical Concepts and Applications in Mechanical Engineering and Mechatronics* (pp. 154–176). Hershey, PA: IGI Global. doi:10.4018/978-1-5225-1639-2.ch008

Henao, J., Poblano-Salas, C. A., Vargas, F., Giraldo-Betancur, A. L., Corona-Castuera, J., & Sotelo-Mazón, O. (2021). Principles and Applications of Thermal Spray Coatings. In S. Roy & G. Bose (Eds.), *Advanced Surface Coating Techniques for Modern Industrial Applications* (pp. 31–70). IGI Global. https://doi.org/10.4018/978-1-7998-4870-7.ch002

Henao, J., & Sotelo, O. (2018). Surface Engineering at High Temperature: Thermal Cycling and Corrosion Resistance. In A. Pakseresht (Ed.), *Production, Properties, and Applications of High Temperature Coatings* (pp. 131–159). Hershey, PA: IGI Global. doi:10.4018/978-1-5225-4194-3.ch006

Hrnčič, M. K., Cör, D., & Knez, Ž. (2021). Supercritical Fluids as a Tool for Green Energy and Chemicals. In L. Chen (Ed.), *Handbook of Research on Advancements in Supercritical Fluids Applications for Sustainable Energy Systems* (pp. 761–791). IGI Global. doi:10.4018/978-1-7998-5796-9.ch021

Ibrahim, O., Erdem, S., & Gurbuz, E. (2021). Studying Physical and Chemical Properties of Graphene Oxide and Reduced Graphene Oxide and Their Applications in Sustainable Building Materials. In R. González-Lezcano (Ed.), *Advancements in Sustainable Architecture and Energy Efficiency* (pp. 221–238). IGI Global. https://doi.org/10.4018/978-1-7998-7023-4.ch010

Ihianle, I. K., Islam, S., Naeem, U., & Ebenuwa, S. H. (2021). Exploiting Patterns of Object Use for Human Activity Recognition. In A. Nwajana & I. Ihianle (Eds.), *Handbook of Research on 5G Networks and Advancements in Computing, Electronics, and Electrical Engineering* (pp. 382–401). IGI Global. https://doi.org/10.4018/978-1-7998-6992-4.ch015

Ijemaru, G. K., Ngharamike, E. T., Oleka, E. U., & Nwajana, A. O. (2021). An Energy-Efficient Model for Opportunistic Data Collection in IoV-Enabled SC Waste Management. In A. Nwajana & I. Ihianle (Eds.), *Handbook of Research on 5G Networks and Advancements in Computing, Electronics, and Electrical Engineering* (pp. 1–19). IGI Global. https://doi.org/10.4018/978-1-7998-6992-4.ch001

Ilori, O. O., Adetan, D. A., & Umoru, L. E. (2017). Effect of Cutting Parameters on the Surface Residual Stress of Face-Milled Pearlitic Ductile Iron. *International Journal of Materials Forming and Machining Processes*, *4*(1), 38–52. doi:10.4018/IJMFMP.2017010103

Imam, M. H., Tasadduq, I. A., Ahmad, A., Aldosari, F., & Khan, H. (2017). Automated Generation of Course Improvement Plans Using Expert System. *International Journal of Quality Assurance in Engineering and Technology Education*, *6*(1), 1–12. doi:10.4018/IJQAETE.2017010101

Injeti, S. K., & Kumar, T. V. (2018). A WDO Framework for Optimal Deployment of DGs and DSCs in a Radial Distribution System Under Daily Load Pattern to Improve Techno-Economic Benefits. *International Journal of Energy Optimization and Engineering*, *7*(2), 1–38. doi:10.4018/IJEOE.2018040101

Ishii, N., Anami, K., & Knisely, C. W. (2018). *Dynamic Stability of Hydraulic Gates and Engineering for Flood Prevention*. Hershey, PA: IGI Global. doi:10.4018/978-1-5225-3079-4

Iwamoto, Y., & Yamaguchi, H. (2021). Application of Supercritical Carbon Dioxide for Solar Water Heater. In L. Chen (Ed.), *Handbook of Research on Advancements in Supercritical Fluids Applications for Sustainable Energy Systems* (pp. 370–387). IGI Global. https://doi.org/10.4018/978-1-7998-5796-9.ch010

Jayapalan, S. (2018). A Review of Chemical Treatments on Natural Fibers-Based Hybrid Composites for Engineering Applications. In K. Kumar & J. Davim (Eds.), *Composites and Advanced Materials for Industrial Applications* (pp. 16–37). Hershey, PA: IGI Global. doi:10.4018/978-1-5225-5216-1.ch002

Kapetanakis, T. N., Vardiambasis, I. O., Ioannidou, M. P., & Konstantaras, A. I. (2021). Modeling Antenna Radiation Using Artificial Intelligence Techniques: The Case of a Circular Loop Antenna. In C. Nikolopoulos (Ed.), *Recent Trends on Electromagnetic Environmental Effects for Aeronautics and Space Applications* (pp. 186–225). IGI Global. https://doi.org/10.4018/978-1-7998-4879-0.ch007

Karkalos, N. E., Markopoulos, A. P., & Dossis, M. F. (2017). Optimal Model Parameters of Inverse Kinematics Solution of a 3R Robotic Manipulator Using ANN Models. *International Journal of Manufacturing, Materials, and Mechanical Engineering*, *7*(3), 20–40. doi:10.4018/IJMMME.2017070102

Kelly, M., Costello, M., Nicholson, G., & O'Connor, J. (2021). The Evolving Integration of BIM Into Built Environment Programmes in a Higher Education Institute. In J. Underwood & M. Shelbourn (Eds.), *Handbook of Research on Driving Transformational Change in the Digital Built Environment* (pp. 294–326). IGI Global. https://doi.org/10.4018/978-1-7998-6600-8.ch012

Kesimal, A., Karaman, K., Cihangir, F., & Ercikdi, B. (2018). Excavatability Assessment of Rock Masses for Geotechnical Studies. In N. Ceryan (Ed.), *Handbook of Research on Trends and Digital Advances in Engineering Geology* (pp. 231–256). Hershey, PA: IGI Global. doi:10.4018/978-1-5225-2709-1.ch006

Knoflacher, H. (2017). The Role of Engineers and Their Tools in the Transport Sector after Paradigm Change: From Assumptions and Extrapolations to Science. In H. Knoflacher & E. Ocalir-Akunal (Eds.), *Engineering Tools and Solutions for Sustainable Transportation Planning* (pp. 1–29). Hershey, PA: IGI Global. doi:10.4018/978-1-5225-2116-7.ch001

Kose, U. (2018). Towards an Intelligent Biomedical Engineering With Nature-Inspired Artificial Intelligence Techniques. In U. Kose, G. Guraksin, & O. Deperlioglu (Eds.), *Nature-Inspired Intelligent Techniques for Solving Biomedical Engineering Problems* (pp. 1–26). Hershey, PA: IGI Global. doi:10.4018/978-1-5225-4769-3.ch001

Kostić, S. (2018). A Review on Enhanced Stability Analyses of Soil Slopes Using Statistical Design. In N. Ceryan (Ed.), *Handbook of Research on Trends and Digital Advances in Engineering Geology* (pp. 446–481). Hershey, PA: IGI Global. doi:10.4018/978-1-5225-2709-1.ch013

Kumar, A., Patil, P. P., & Prajapati, Y. K. (2018). *Advanced Numerical Simulations in Mechanical Engineering*. Hershey, PA: IGI Global. doi:10.4018/978-1-5225-3722-9

Kumar, G. R., Rajyalakshmi, G., & Manupati, V. K. (2017). Surface Micro Patterning of Aluminium Reinforced Composite through Laser Peening. *International Journal of Manufacturing, Materials, and Mechanical Engineering*, *7*(4), 15–27. doi:10.4018/IJMMME.2017100102

Kumar, N., Basu, D. N., & Chen, L. (2021). Effect of Flow Acceleration and Buoyancy on Thermalhydraulics of sCO2 in Mini/Micro-Channel. In L. Chen (Ed.), *Handbook of Research on Advancements in Supercritical Fluids Applications for Sustainable Energy Systems* (pp. 161–182). IGI Global. doi:10.4018/978-1-7998-5796-9.ch005

Kumari, N., & Kumar, K. (2018). Fabrication of Orthotic Calipers With Epoxy-Based Green Composite. In K. Kumar & J. Davim (Eds.), *Composites and Advanced Materials for Industrial Applications* (pp. 157–176). Hershey, PA: IGI Global. doi:10.4018/978-1-5225-5216-1.ch008

Kuppusamy, R. R. (2018). Development of Aerospace Composite Structures Through Vacuum-Enhanced Resin Transfer Moulding Technology (VERTMTy): Vacuum-Enhanced Resin Transfer Moulding. In K. Kumar & J. Davim (Eds.), *Composites and Advanced Materials for Industrial Applications* (pp. 99–111). Hershey, PA: IGI Global. doi:10.4018/978-1-5225-5216-1.ch005

Kurganov, V. A., Zeigarnik, Y. A., & Maslakova, I. V. (2021). Normal and Deteriorated Heat Transfer Under Heating Turbulent Supercritical Pressure Coolants Flows in Round Tubes. In L. Chen (Ed.), *Handbook of Research on Advancements in Supercritical Fluids Applications for Sustainable Energy Systems* (pp. 494–532). IGI Global. https://doi.org/10.4018/978-1-7998-5796-9.ch014

Li, H., & Zhang, Y. (2021). Heat Transfer and Fluid Flow Modeling for Supercritical Fluids in Advanced Energy Systems. In L. Chen (Ed.), *Handbook of Research on Advancements in Supercritical Fluids Applications for Sustainable Energy Systems* (pp. 388–422). IGI Global. https://doi.org/10.4018/978-1-7998-5796-9.ch011

Loy, J., Howell, S., & Cooper, R. (2017). Engineering Teams: Supporting Diversity in Engineering Education. In M. Gray & K. Thomas (Eds.), *Strategies for Increasing Diversity in Engineering Majors and Careers* (pp. 106–129). Hershey, PA: IGI Global. doi:10.4018/978-1-5225-2212-6.ch006

Macher, G., Armengaud, E., Kreiner, C., Brenner, E., Schmittner, C., Ma, Z., ... Krammer, M. (2018). Integration of Security in the Development Lifecycle of Dependable Automotive CPS. In N. Druml, A. Genser, A. Krieg, M. Menghin, & A. Hoeller (Eds.), *Solutions for Cyber-Physical Systems Ubiquity* (pp. 383–423). Hershey, PA: IGI Global. doi:10.4018/978-1-5225-2845-6.ch015

Madhu, M. N., Singh, J. G., Mohan, V., & Ongsakul, W. (2021). Transmission Risk Optimization in Interconnected Systems: Risk-Adjusted Available Transfer Capability. In P. Vasant, G. Weber, & W. Punurai (Ed.), *Research Advancements in Smart Technology, Optimization, and Renewable Energy* (pp. 183-199). IGI Global. https://doi.org/10.4018/978-1-7998-3970-5.ch010

Mahendramani, G., & Lakshmana Swamy, N. (2018). Effect of Weld Groove Area on Distortion of Butt Welded Joints in Submerged Arc Welding. *International Journal of Manufacturing, Materials, and Mechanical Engineering*, *8*(2), 33–44. doi:10.4018/IJMMME.2018040103

Makropoulos, G., Koumaras, H., Setaki, F., Filis, K., Lutz, T., Montowtt, P., Tomaszewski, L., Dybiec, P., & Järvet, T. (2021). 5G and Unmanned Aerial Vehicles (UAVs) Use Cases: Analysis of the Ecosystem, Architecture, and Applications. In A. Nwajana & I. Ihianle (Eds.), *Handbook of Research on 5G Networks and Advancements in Computing, Electronics, and Electrical Engineering* (pp. 36–69). IGI Global. https://doi.org/10.4018/978-1-7998-6992-4.ch003

Meric, E. M., Erdem, S., & Gurbuz, E. (2021). Application of Phase Change Materials in Construction Materials for Thermal Energy Storage Systems in Buildings. In R. González-Lezcano (Ed.), *Advancements in Sustainable Architecture and Energy Efficiency* (pp. 1–20). IGI Global. https://doi.org/10.4018/978-1-7998-7023-4.ch001

Mihret, E. T., & Yitayih, K. A. (2021). Operation of VANET Communications: The Convergence of UAV System With LTE/4G and WAVE Technologies. *International Journal of Smart Vehicles and Smart Transportation*, *4*(1), 29–51. https://doi.org/10.4018/IJSVST.2021010103

Mir, M. A., Bhat, B. A., Sheikh, B. A., Rather, G. A., Mehraj, S., & Mir, W. R. (2021). Nanomedicine in Human Health Therapeutics and Drug Delivery: Nanobiotechnology and Nanobiomedicine. In M. Bhat, I. Wani, & S. Ashraf (Eds.), *Applications of Nanomaterials in Agriculture, Food Science, and Medicine* (pp. 229–251). IGI Global. doi:10.4018/978-1-7998-5563-7.ch013

Mohammadzadeh, S., & Kim, Y. (2017). Nonlinear System Identification of Smart Buildings. In P. Samui, S. Chakraborty, & D. Kim (Eds.), *Modeling and Simulation Techniques in Structural Engineering* (pp. 328–347). Hershey, PA: IGI Global. doi:10.4018/978-1-5225-0588-4.ch011

Molina, G. J., Aktaruzzaman, F., Soloiu, V., & Rahman, M. (2017). Design and Testing of a Jet-Impingement Instrument to Study Surface-Modification Effects by Nanofluids. *International Journal of Surface Engineering and Interdisciplinary Materials Science*, *5*(2), 43–61. doi:10.4018/IJSEIMS.2017070104

Moreno-Rangel, A., & Carrillo, G. (2021). Energy-Efficient Homes: A Heaven for Respiratory Illnesses. In R. González-Lezcano (Ed.), *Advancements in Sustainable Architecture and Energy Efficiency* (pp. 49–71). IGI Global. https://doi.org/10.4018/978-1-7998-7023-4.ch003

Msomi, V., & Jantjies, B. T. (2021). Correlative Analysis Between Tensile Properties and Tool Rotational Speeds of Friction Stir Welded Similar Aluminium Alloy Joints. *International Journal of Surface Engineering and Interdisciplinary Materials Science*, *9*(2), 58–78. https://doi.org/10.4018/IJSEIMS.2021070104

Muigai, M. N., Mwema, F. M., Akinlabi, E. T., & Obiko, J. O. (2021). Surface Engineering of Materials Through Weld-Based Technologies: An Overview. In S. Roy & G. Bose (Eds.), *Advanced Surface Coating Techniques for Modern Industrial Applications* (pp. 247–260). IGI Global. doi:10.4018/978-1-7998-4870-7.ch011

Mukherjee, A., Saeed, R. A., Dutta, S., & Naskar, M. K. (2017). Fault Tracking Framework for Software-Defined Networking (SDN). In C. Singhal & S. De (Eds.), *Resource Allocation in Next-Generation Broadband Wireless Access Networks* (pp. 247–272). Hershey, PA: IGI Global. doi:10.4018/978-1-5225-2023-8.ch011

Mukhopadhyay, A., Barman, T. K., & Sahoo, P. (2018). Electroless Nickel Coatings for High Temperature Applications. In K. Kumar & J. Davim (Eds.), *Composites and Advanced Materials for Industrial Applications* (pp. 297–331). Hershey, PA: IGI Global. doi:10.4018/978-1-5225-5216-1.ch013

Mwema, F. M., & Wambua, J. M. (2022). Machining of Poly Methyl Methacrylate (PMMA) and Other Olymeric Materials: A Review. In K. Kumar, B. Babu, & J. Davim (Eds.), *Handbook of Research on Advancements in the Processing, Characterization, and Application of Lightweight Materials* (pp. 363–379). IGI Global. https://doi.org/10.4018/978-1-7998-7864-3.ch016

Mykhailyshyn, R., Savkiv, V., Boyko, I., Prada, E., & Virgala, I. (2021). Substantiation of Parameters of Friction Elements of Bernoulli Grippers With a Cylindrical Nozzle. *International Journal of Manufacturing, Materials, and Mechanical Engineering, 11*(2), 17–39. https://doi.org/10.4018/IJMMME.2021040102

Náprstek, J., & Fischer, C. (2017). Dynamic Stability and Post-Critical Processes of Slender Auto-Parametric Systems. In V. Plevris, G. Kremmyda, & Y. Fahjan (Eds.), *Performance-Based Seismic Design of Concrete Structures and Infrastructures* (pp. 128–171). Hershey, PA: IGI Global. doi:10.4018/978-1-5225-2089-4.ch006

Nautiyal, L., Shivach, P., & Ram, M. (2018). Optimal Designs by Means of Genetic Algorithms. In M. Ram & J. Davim (Eds.), *Soft Computing Techniques and Applications in Mechanical Engineering* (pp. 151–161). Hershey, PA: IGI Global. doi:10.4018/978-1-5225-3035-0.ch007

Nazir, R. (2017). Advanced Nanomaterials for Water Engineering and Treatment: Nano-Metal Oxides and Their Nanocomposites. In T. Saleh (Ed.), *Advanced Nanomaterials for Water Engineering, Treatment, and Hydraulics* (pp. 84–126). Hershey, PA: IGI Global. doi:10.4018/978-1-5225-2136-5.ch005

Nikolopoulos, C. D. (2021). Recent Advances on Measuring and Modeling ELF-Radiated Emissions for Space Applications. In C. Nikolopoulos (Ed.), *Recent Trends on Electromagnetic Environmental Effects for Aeronautics and Space Applications* (pp. 1–38). IGI Global. https://doi.org/10.4018/978-1-7998-4879-0.ch001

Nogueira, A. F., Ribeiro, J. C., Fernández de Vega, F., & Zenha-Rela, M. A. (2018). Evolutionary Approaches to Test Data Generation for Object-Oriented Software: Overview of Techniques and Tools. In M. Khosrow-Pour, D.B.A. (Ed.), Incorporating Nature-Inspired Paradigms in Computational Applications (pp. 162-194). Hershey, PA: IGI Global. https://doi.org/ doi:10.4018/978-1-5225-5020-4.ch006

Nwajana, A. O., Obi, E. R., Ijemaru, G. K., Oleka, E. U., & Anthony, D. C. (2021). Fundamentals of RF/Microwave Bandpass Filter Design. In A. Nwajana & I. Ihianle (Eds.), *Handbook of Research on 5G Networks and Advancements in Computing, Electronics, and Electrical Engineering* (pp. 149–164). IGI Global. https://doi.org/10.4018/978-1-7998-6992-4.ch005

Ogbodo, E. A. (2021). Comparative Study of Transmission Line Junction vs. Asynchronously Coupled Junction Diplexers. In A. Nwajana & I. Ihianle (Eds.), *Handbook of Research on 5G Networks and Advancements in Computing, Electronics, and Electrical Engineering* (pp. 326–336). IGI Global. https://doi.org/10.4018/978-1-7998-6992-4.ch013

Orosa, J. A., Vergara, D., Fraguela, F., & Masdías-Bonome, A. (2021). Statistical Understanding and Optimization of Building Energy Consumption and Climate Change Consequences. In R. González-Lezcano (Ed.), *Advancements in Sustainable Architecture and Energy Efficiency* (pp. 195–220). IGI Global. https://doi.org/10.4018/978-1-7998-7023-4.ch009

Osho, M. B. (2018). Industrial Enzyme Technology: Potential Applications. In S. Bharati & P. Chaurasia (Eds.), *Research Advancements in Pharmaceutical, Nutritional, and Industrial Enzymology* (pp. 375–394). Hershey, PA: IGI Global. doi:10.4018/978-1-5225-5237-6.ch017

Ouadi, A., & Zitouni, A. (2021). Phasor Measurement Improvement Using Digital Filter in a Smart Grid. In A. Recioui & H. Bentarzi (Eds.), *Optimizing and Measuring Smart Grid Operation and Control* (pp. 100–117). IGI Global. https://doi.org/10.4018/978-1-7998-4027-5.ch005

Padmaja, P., & Marutheswar, G. (2017). Certain Investigation on Secured Data Transmission in Wireless Sensor Networks. *International Journal of Mobile Computing and Multimedia Communications*, *8*(1), 48–61. doi:10.4018/IJMCMC.2017010104

Palmer, S., & Hall, W. (2017). An Evaluation of Group Work in First-Year Engineering Design Education. In R. Tucker (Ed.), *Collaboration and Student Engagement in Design Education* (pp. 145–168). Hershey, PA: IGI Global. doi:10.4018/978-1-5225-0726-0.ch007

Panchenko, V. (2021). Prospects for Energy Supply of the Arctic Zone Objects of Russia Using Frost-Resistant Solar Modules. In P. Vasant, G. Weber, & W. Punurai (Eds.), *Research Advancements in Smart Technology, Optimization, and Renewable Energy* (pp. 149-169). IGI Global. https://doi.org/10.4018/978-1-7998-3970-5.ch008

Panchenko, V. (2021). Photovoltaic Thermal Module With Paraboloid Type Solar Concentrators. *International Journal of Energy Optimization and Engineering*, *10*(2), 1–23. https://doi.org/10.4018/IJEOE.2021040101

Pandey, K., & Datta, S. (2021). Dry Machining of Inconel 825 Superalloys: Performance of Tool Inserts (Carbide, Cermet, and SiAlON). *International Journal of Manufacturing, Materials, and Mechanical Engineering*, *11*(4), 26–39. doi:10.4018/IJMMME.2021100102

Panneer, R. (2017). Effect of Composition of Fibers on Properties of Hybrid Composites. *International Journal of Manufacturing, Materials, and Mechanical Engineering*, *7*(4), 28–43. doi:10.4018/IJMMME.2017100103

Pany, C. (2021). Estimation of Correct Long-Seam Mismatch Using FEA to Compare the Measured Strain in a Non-Destructive Testing of a Pressurant Tank: A Reverse Problem. *International Journal of Smart Vehicles and Smart Transportation*, *4*(1), 16–28. doi:10.4018/IJSVST.2021010102

Paul, S., & Roy, P. (2018). Optimal Design of Power System Stabilizer Using a Novel Evolutionary Algorithm. *International Journal of Energy Optimization and Engineering*, *7*(3), 24–46. doi:10.4018/IJEOE.2018070102

Paul, S., & Roy, P. K. (2021). Oppositional Differential Search Algorithm for the Optimal Tuning of Both Single Input and Dual Input Power System Stabilizer. In P. Vasant, G. Weber, & W. Punurai (Eds.), *Research Advancements in Smart Technology, Optimization, and Renewable Energy* (pp. 256-282). IGI Global. https://doi.org/10.4018/978-1-7998-3970-5.ch013

Pavaloiu, A. (2018). Artificial Intelligence Ethics in Biomedical-Engineering-Oriented Problems. In U. Kose, G. Guraksin, & O. Deperlioglu (Eds.), *Nature-Inspired Intelligent Techniques for Solving Biomedical Engineering Problems* (pp. 219–231). Hershey, PA: IGI Global. doi:10.4018/978-1-5225-4769-3.ch010

Pioro, I., Mahdi, M., & Popov, R. (2017). Application of Supercritical Pressures in Power Engineering. In L. Chen & Y. Iwamoto (Eds.), *Advanced Applications of Supercritical Fluids in Energy Systems* (pp. 404–457). Hershey, PA: IGI Global. doi:10.4018/978-1-5225-2047-4.ch013

Plaksina, T., & Gildin, E. (2017). Rigorous Integrated Evolutionary Workflow for Optimal Exploitation of Unconventional Gas Assets. *International Journal of Energy Optimization and Engineering*, *6*(1), 101–122. doi:10.4018/IJEOE.2017010106

Popat, J., Kakadiya, H., Tak, L., Singh, N. K., Majeed, M. A., & Mahajan, V. (2021). Reliability of Smart Grid Including Cyber Impact: A Case Study. In R. Singh, A. Singh, A. Dwivedi, & P. Nagabhushan (Eds.), *Computational Methodologies for Electrical and Electronics Engineers* (pp. 163–174). IGI Global. https://doi.org/10.4018/978-1-7998-3327-7.ch013

Quiza, R., La Fé-Perdomo, I., Rivas, M., & Ramtahalsing, V. (2021). Triple Bottom Line-Focused Optimization of Oblique Turning Processes Based on Hybrid Modeling: A Study Case on AISI 1045 Steel Turning. In L. Burstein (Ed.), *Handbook of Research on Advancements in Manufacturing, Materials, and Mechanical Engineering* (pp. 215–241). IGI Global. https://doi.org/10.4018/978-1-7998-4939-1.ch010

Rahmani, M. K. (2022). Blockchain Technology: Principles and Algorithms. In S. Khan, M. Syed, R. Hammad, & A. Bushager (Eds.), *Blockchain Technology and Computational Excellence for Society 5.0* (pp. 16–27). IGI Global. https://doi.org/10.4018/978-1-7998-8382-1.ch002

Ramdani, N., & Azibi, M. (2018). Polymer Composite Materials for Microelectronics Packaging Applications: Composites for Microelectronics Packaging. In K. Kumar & J. Davim (Eds.), *Composites and Advanced Materials for Industrial Applications* (pp. 177–211). Hershey, PA: IGI Global. doi:10.4018/978-1-5225-5216-1.ch009

Ramesh, M., Garg, R., & Subrahmanyam, G. V. (2017). Investigation of Influence of Quenching and Annealing on the Plane Fracture Toughness and Brittle to Ductile Transition Temperature of the Zinc Coated Structural Steel Materials. *International Journal of Surface Engineering and Interdisciplinary Materials Science*, *5*(2), 33–42. doi:10.4018/IJSEIMS.2017070103

Robinson, J., & Beneroso, D. (2022). Project-Based Learning in Chemical Engineering: Curriculum and Assessment, Culture and Learning Spaces. In A. Alves & N. van Hattum-Janssen (Eds.), *Training Engineering Students for Modern Technological Advancement* (pp. 1–19). IGI Global. https://doi.org/10.4018/978-1-7998-8816-1.ch001

Rondon, B. (2021). Experimental Characterization of Admittance Meter With Crude Oil Emulsions. *International Journal of Electronics, Communications, and Measurement Engineering*, *10*(2), 51–59. https://doi.org/10.4018/IJECME.2021070104

Rudolf, S., Biryuk, V. V., & Volov, V. (2018). Vortex Effect, Vortex Power: Technology of Vortex Power Engineering. In V. Kharchenko & P. Vasant (Eds.), *Handbook of Research on Renewable Energy and Electric Resources for Sustainable Rural Development* (pp. 500–533). Hershey, PA: IGI Global. doi:10.4018/978-1-5225-3867-7.ch021

Sah, A., Bhadula, S. J., Dumka, A., & Rawat, S. (2018). A Software Engineering Perspective for Development of Enterprise Applications. In A. Elçi (Ed.), *Handbook of Research on Contemporary Perspectives on Web-Based Systems* (pp. 1–23). Hershey, PA: IGI Global. doi:10.4018/978-1-5225-5384-7.ch001

Sahli, Y., Zitouni, B., & Hocine, B. M. (2021). Three-Dimensional Numerical Study of Overheating of Two Intermediate Temperature P-AS-SOFC Geometrical Configurations. In G. Badea, R. Felseghi, & I. Aşchilean (Eds.), *Hydrogen Fuel Cell Technology for Stationary Applications* (pp. 186–222). IGI Global. https://doi.org/10.4018/978-1-7998-4945-2.ch008

Sahoo, P., & Roy, S. (2017). Tribological Behavior of Electroless Ni-P, Ni-P-W and Ni-P-Cu Coatings: A Comparison. *International Journal of Surface Engineering and Interdisciplinary Materials Science*, *5*(1), 1–15. doi:10.4018/IJSEIMS.2017010101

Sahoo, S. (2018). Laminated Composite Hypar Shells as Roofing Units: Static and Dynamic Behavior. In K. Kumar & J. Davim (Eds.), *Composites and Advanced Materials for Industrial Applications* (pp. 249–269). Hershey, PA: IGI Global. doi:10.4018/978-1-5225-5216-1.ch011

Sahu, H., & Hungyo, M. (2018). Introduction to SDN and NFV. In A. Dumka (Ed.), *Innovations in Software-Defined Networking and Network Functions Virtualization* (pp. 1–25). Hershey, PA: IGI Global. doi:10.4018/978-1-5225-3640-6.ch001

Salem, A. M., & Shmelova, T. (2018). Intelligent Expert Decision Support Systems: Methodologies, Applications, and Challenges. In T. Shmelova, Y. Sikirda, N. Rizun, A. Salem, & Y. Kovalyov (Eds.), *Socio-Technical Decision Support in Air Navigation Systems: Emerging Research and Opportunities* (pp. 215–242). Hershey, PA: IGI Global. doi:10.4018/978-1-5225-3108-1.ch007

Samal, M. (2017). FE Analysis and Experimental Investigation of Cracked and Un-Cracked Thin-Walled Tubular Components to Evaluate Mechanical and Fracture Properties. In P. Samui, S. Chakraborty, & D. Kim (Eds.), *Modeling and Simulation Techniques in Structural Engineering* (pp. 266–293). Hershey, PA: IGI Global. doi:10.4018/978-1-5225-0588-4.ch009

Samal, M., & Balakrishnan, K. (2017). Experiments on a Ring Tension Setup and FE Analysis to Evaluate Transverse Mechanical Properties of Tubular Components. In P. Samui, S. Chakraborty, & D. Kim (Eds.), *Modeling and Simulation Techniques in Structural Engineering* (pp. 91–115). Hershey, PA: IGI Global. doi:10.4018/978-1-5225-0588-4.ch004

Samarasinghe, D. A., & Wood, E. (2021). Innovative Digital Technologies. In J. Underwood & M. Shelbourn (Eds.), *Handbook of Research on Driving Transformational Change in the Digital Built Environment* (pp. 142–163). IGI Global. https://doi.org/10.4018/978-1-7998-6600-8.ch006

Sawant, S. (2018). Deep Learning and Biomedical Engineering. In U. Kose, G. Guraksin, & O. Deperlioglu (Eds.), *Nature-Inspired Intelligent Techniques for Solving Biomedical Engineering Problems* (pp. 283–296). Hershey, PA: IGI Global. doi:10.4018/978-1-5225-4769-3.ch014

Schulenberg, T. (2021). Energy Conversion Using the Supercritical Steam Cycle. In L. Chen (Ed.), *Handbook of Research on Advancements in Supercritical Fluids Applications for Sustainable Energy Systems* (pp. 659–681). IGI Global. doi:10.4018/978-1-7998-5796-9.ch018

Sezgin, H., & Berkalp, O. B. (2018). Textile-Reinforced Composites for the Automotive Industry. In K. Kumar & J. Davim (Eds.), *Composites and Advanced Materials for Industrial Applications* (pp. 129–156). Hershey, PA: IGI Global. doi:10.4018/978-1-5225-5216-1.ch007

Shaaban, A. A., & Shehata, O. M. (2021). Combining Response Surface Method and Metaheuristic Algorithms for Optimizing SPIF Process. *International Journal of Manufacturing, Materials, and Mechanical Engineering*, *11*(4), 1–25. https://doi.org/10.4018/IJMMME.2021100101

Shafaati Shemami, M., & Sefid, M. (2022). Implementation and Demonstration of Electric Vehicle-to-Home (V2H) Application: A Case Study. In M. Alam, R. Pillai, & N. Murugesan (Eds.), *Developing Charging Infrastructure and Technologies for Electric Vehicles* (pp. 268–293). IGI Global. https://doi.org/10.4018/978-1-7998-6858-3.ch015

Shah, M. Z., Gazder, U., Bhatti, M. S., & Hussain, M. (2018). Comparative Performance Evaluation of Effects of Modifier in Asphaltic Concrete Mix. *International Journal of Strategic Engineering*, *1*(2), 13–25. doi:10.4018/IJoSE.2018070102

Sharma, N., & Kumar, K. (2018). Fabrication of Porous NiTi Alloy Using Organic Binders. In K. Kumar & J. Davim (Eds.), *Composites and Advanced Materials for Industrial Applications* (pp. 38–62). Hershey, PA: IGI Global. doi:10.4018/978-1-5225-5216-1.ch003

Shivach, P., Nautiyal, L., & Ram, M. (2018). Applying Multi-Objective Optimization Algorithms to Mechanical Engineering. In M. Ram & J. Davim (Eds.), *Soft Computing Techniques and Applications in Mechanical Engineering* (pp. 287–301). Hershey, PA: IGI Global. doi:10.4018/978-1-5225-3035-0.ch014

Shmelova, T. (2018). Stochastic Methods for Estimation and Problem Solving in Engineering: Stochastic Methods of Decision Making in Aviation. In S. Kadry (Ed.), *Stochastic Methods for Estimation and Problem Solving in Engineering* (pp. 139–160). Hershey, PA: IGI Global. doi:10.4018/978-1-5225-5045-7.ch006

Siero González, L. R., & Romo Vázquez, A. (2017). Didactic Sequences Teaching Mathematics for Engineers With Focus on Differential Equations. In M. Ramírez-Montoya (Ed.), *Handbook of Research on Driving STEM Learning With Educational Technologies* (pp. 129–151). Hershey, PA: IGI Global. doi:10.4018/978-1-5225-2026-9.ch007

Sim, M. S., You, K. Y., Esa, F., & Chan, Y. L. (2021). Nanostructured Electromagnetic Metamaterials for Sensing Applications. In M. Bhat, I. Wani, & S. Ashraf (Eds.), *Applications of Nanomaterials in Agriculture, Food Science, and Medicine* (pp. 141–164). IGI Global. https://doi.org/10.4018/978-1-7998-5563-7.ch009

Singh, R., & Dutta, S. (2018). Visible Light Active Nanocomposites for Photocatalytic Applications. In K. Kumar & J. Davim (Eds.), *Composites and Advanced Materials for Industrial Applications* (pp. 270–296). Hershey, PA: IGI Global. doi:10.4018/978-1-5225-5216-1.ch012

Skripov, P. V., Yampol'skiy, A. D., & Rutin, S. B. (2021). High-Power Heat Transfer in Supercritical Fluids: Microscale Times and Sizes. In L. Chen (Ed.), *Handbook of Research on Advancements in Supercritical Fluids Applications for Sustainable Energy Systems* (pp. 424–450). IGI Global. https://doi.org/10.4018/978-1-7998-5796-9.ch012

Sözbilir, H., Özkaymak, Ç., Uzel, B., & Sümer, Ö. (2018). Criteria for Surface Rupture Microzonation of Active Faults for Earthquake Hazards in Urban Areas. In N. Ceryan (Ed.), *Handbook of Research on Trends and Digital Advances in Engineering Geology* (pp. 187–230). Hershey, PA: IGI Global. doi:10.4018/978-1-5225-2709-1.ch005

Stanciu, I. (2018). Stochastic Methods in Microsystems Engineering. In S. Kadry (Ed.), *Stochastic Methods for Estimation and Problem Solving in Engineering* (pp. 161–176). Hershey, PA: IGI Global. doi:10.4018/978-1-5225-5045-7.ch007

Strebkov, D., Nekrasov, A., Trubnikov, V., & Nekrasov, A. (2018). Single-Wire Resonant Electric Power Systems for Renewable-Based Electric Grid. In V. Kharchenko & P. Vasant (Eds.), *Handbook of Research on Renewable Energy and Electric Resources for Sustainable Rural Development* (pp. 449–474). Hershey, PA: IGI Global. doi:10.4018/978-1-5225-3867-7.ch019

Sukhyy, K., Belyanovskaya, E., & Sukhyy, M. (2021). *Basic Principles for Substantiation of Working Pair Choice*. IGI Global. doi:10.4018/978-1-7998-4432-7.ch002

Suri, M. S., & Kaliyaperumal, D. (2022). Extension of Aspiration Level Model for Optimal Planning of Fast Charging Stations. In A. Fekik & N. Benamrouche (Eds.), *Modeling and Control of Static Converters for Hybrid Storage Systems* (pp. 91–106). IGI Global. https://doi.org/10.4018/978-1-7998-7447-8.ch004

Tallet, E., Gledson, B., Rogage, K., Thompson, A., & Wiggett, D. (2021). Digitally-Enabled Design Management. In J. Underwood & M. Shelbourn (Eds.), *Handbook of Research on Driving Transformational Change in the Digital Built Environment* (pp. 63–89). IGI Global. https://doi.org/10.4018/978-1-7998-6600-8.ch003

Terki, A., & Boubertakh, H. (2021). A New Hybrid Binary-Real Coded Cuckoo Search and Tabu Search Algorithm for Solving the Unit-Commitment Problem. *International Journal of Energy Optimization and Engineering, 10*(2), 104–119. https://doi.org/10.4018/IJEOE.2021040105

Tüdeş, Ş., Kumlu, K. B., & Ceryan, S. (2018). Integration Between Urban Planning and Natural Hazards For Resilient City. In N. Ceryan (Ed.), *Handbook of Research on Trends and Digital Advances in Engineering Geology* (pp. 591–630). Hershey, PA: IGI Global. doi:10.4018/978-1-5225-2709-1.ch017

Ulamis, K. (2018). Soil Liquefaction Assessment by Anisotropic Cyclic Triaxial Test. In N. Ceryan (Ed.), *Handbook of Research on Trends and Digital Advances in Engineering Geology* (pp. 631–664). Hershey, PA: IGI Global. doi:10.4018/978-1-5225-2709-1.ch018

Valente, M., & Milani, G. (2017). Seismic Assessment and Retrofitting of an Under-Designed RC Frame Through a Displacement-Based Approach. In V. Plevris, G. Kremmyda, & Y. Fahjan (Eds.), *Performance-Based Seismic Design of Concrete Structures and Infrastructures* (pp. 36–58). Hershey, PA: IGI Global. doi:10.4018/978-1-5225-2089-4.ch002

Vargas-Bernal, R. (2021). Advances in Electromagnetic Environmental Shielding for Aeronautics and Space Applications. In C. Nikolopoulos (Ed.), *Recent Trends on Electromagnetic Environmental Effects for Aeronautics and Space Applications* (pp. 80–96). IGI Global. https://doi.org/10.4018/978-1-7998-4879-0.ch003

Vasant, P. (2018). A General Medical Diagnosis System Formed by Artificial Neural Networks and Swarm Intelligence Techniques. In U. Kose, G. Guraksin, & O. Deperlioglu (Eds.), *Nature-Inspired Intelligent Techniques for Solving Biomedical Engineering Problems* (pp. 130–145). Hershey, PA: IGI Global. doi:10.4018/978-1-5225-4769-3.ch006

Verner, C. M., & Sarwar, D. (2021). Avoiding Project Failure and Achieving Project Success in NHS IT System Projects in the United Kingdom. *International Journal of Strategic Engineering*, *4*(1), 33–54. https://doi.org/10.4018/IJoSE.2021010103

Verrollot, J., Tolonen, A., Harkonen, J., & Haapasalo, H. J. (2018). Challenges and Enablers for Rapid Product Development. *International Journal of Applied Industrial Engineering*, *5*(1), 25–49. doi:10.4018/IJAIE.2018010102

Wan, A. C., Zulu, S. L., & Khosrow-Shahi, F. (2021). Industry Views on BIM for Site Safety in Hong Kong. In J. Underwood & M. Shelbourn (Eds.), *Handbook of Research on Driving Transformational Change in the Digital Built Environment* (pp. 120–140). IGI Global. https://doi.org/10.4018/978-1-7998-6600-8.ch005

Yardimci, A. G., & Karpuz, C. (2018). Fuzzy Rock Mass Rating: Soft-Computing-Aided Preliminary Stability Analysis of Weak Rock Slopes. In N. Ceryan (Ed.), *Handbook of Research on Trends and Digital Advances in Engineering Geology* (pp. 97–131). Hershey, PA: IGI Global. doi:10.4018/978-1-5225-2709-1.ch003

You, K. Y. (2021). Development Electronic Design Automation for RF/Microwave Antenna Using MATLAB GUI. In A. Nwajana & I. Ihianle (Eds.), *Handbook of Research on 5G Networks and Advancements in Computing, Electronics, and Electrical Engineering* (pp. 70–148). IGI Global. https://doi.org/10.4018/978-1-7998-6992-4.ch004

Yousefi, Y., Gratton, P., & Sarwar, D. (2021). Investigating the Opportunities to Improve the Thermal Performance of a Case Study Building in London. *International Journal of Strategic Engineering*, *4*(1), 1–18. https://doi.org/10.4018/IJoSE.2021010101

Zindani, D., & Kumar, K. (2018). Industrial Applications of Polymer Composite Materials. In K. Kumar & J. Davim (Eds.), *Composites and Advanced Materials for Industrial Applications* (pp. 1–15). Hershey, PA: IGI Global. doi:10.4018/978-1-5225-5216-1.ch001

Zindani, D., Maity, S. R., & Bhowmik, S. (2018). A Decision-Making Approach for Material Selection of Polymeric Composite Bumper Beam. In K. Kumar & J. Davim (Eds.), *Composites and Advanced Materials for Industrial Applications* (pp. 112–128). Hershey, PA: IGI Global. doi:10.4018/978-1-5225-5216-1.ch006

# About the Contributors

**Ayşe Aslı Yılmaz** is an Assistant Professorand works as chair of the Pilot Training Department at Civil Aviation School of Atilim University. She earned her Ph.D. in Business Administration (English program) from the Social Sciences Institute at Atılım University in 2022, focusing on her dissertation titled "The Value Added by Digital Transformation Criteria to Intellectual Capital Components: Network Analysis on Esenboğa Airport." With a background in Peace Support Operations (PSO), Dr. Yılmaz has served as a course director and instructor on various topics including Cultural Awareness in Multinational Environments, Strategic Communication, Civil-Military Cooperation, and Gender Awareness in PSOs at Partnership for Peace Training Centre, as well as several UN and NATO affiliated agencies and training centres since 2011. Her research interests lie in social networks, intellectual capital, and digital transformation, Aviation Psychology, and Fatigue Risk Management. Dr. Yılmaz has contributed significantly to the field through her writings of book chapters and articles in management and organization domains. She teaches a range of classes including Airport Management, Airport Master Planning, International Aviation Law, and Air Traffic Rules and Regulations. Additionally, she instructs courses on Crew Resource Management (CRM) and Accident Investigation Analysis. Dr. Yılmaz is also a member of the European Association for Aviation Psychology (EAAP) and has UAV (Unmanned Aerial Vehicle) experience.

***

**Hüseyin Önder Aldemir** is an Assistant Professor at Özyeğin University, Faculty of Aviation and Aeronautical Sciences. He obtained his B.S. from Army School, Systems Engineering in 2000, M.S. in Mechanical Engineering from The Pennsylvania State University (PennState) in 2005, and Ph.D. in Civil Aviation Management from Anadolu University in 2018. He was accepted as a member of Air Transport and Regional Development (ATARD), which is a sub-organization of European Cooperation in Science and Technology (COST). Within this framework, he participated two Ph.D. schools which were held at University of Bergamo,

ITALY in 2017 and Poznan University of Economics and Business, POLAND in 2018. Dr. ALDEMİR accomplished Unmanned Aerial Vehicle (UAV) Systems pilot training in ISRAEL in 2008 and he has over 2500-hour flight with different types of UAV systems as UAS pilot and UAS test/instructor pilot. His researches focus on strategic management, competition strategies, airline management, and Unmanned Aerial Vehicle Systems (UAS).

**Caner Asbaş** is founder and CEO of Hamle RF Elektrik ve Elektronik A.Ş. which is a company operating in defense, telecommunication, avionics and automotive areas. He received his undergraduate and MSc degrees from İhsan Doğramacı Bilkent University in Electrical and Electronics Engineering and PhD degree from Atılım University in Business Administration He also has undergraduate degrees in International Relations from Anadolu University, Sociology from Anadolu University, and Public Finance from Anadolu University. He is proficient in C, C++, C#, Java, MSSQL and Python programming languages. His research focuses on social networks, implementation of artificial intelligence methods to social networks, cyberthreats and digitlization.

**Tugba Erhan** (Ph.D.) is an associate professor of business at Suleyman Demirel University. She earned her BA in English Language Teaching from Uludag University and PhD in Business from Suleyman Demirel University. She conducts positive organizational behavior research mainly focusing on job approaches of employees, sense of calling, and decent work. She teaches Organizational Behavior, Career Management, Contemporary Management Approaches, Perception and Reputation Management at undergraduate and graduate levels. She is the language editor of Sosyal Mucit Review ve Journal of Aviation Research (JAR), and the assistant editor of European Journal of Digital Economy Research.

**Inan Eryilmaz** was born in Ankara and completed his primary and secondary education in Ankara. Between 2005 and 2020, he did his duty in the Land Aviation Units of the Turkish Armed Forces on aircraft maintenance, quality control and maintenance planning. During his service in the Turkish Armed Forces, he received his bachelor's degree from Manisa Celal Bayar University, Department of Business Administration (2009), master's degree from the Department of Management and Organization (2012) and doctorate degree from the Department of Business Administration (2017). In 2020, he became an Associate Professor in the field of Management and Strategy. Between 2019-2021, he worked at Maltepe University Aircraft Technology program. He continues to work as a faculty member at Süleyman Demirel University, School of Civil Aviation, Department of Aviation Management (2021-.). Dr. Eryılmaz, who continues his research on leadership, motivation, organizational strength, human factors in aviation and team resource management, is the editor-in-chief of the Journal of Aviation Research (JAR) and area editor of the Journal of Aviation.

**Tereza Raquel Merlo** is a researcher, professor, practitioner, and consultant working for the University of North Texas, where she teaches courses in Knowledge Management Tools and Technologies. She is CEO of Merlo Management Consulting, providing solutions to businesses around the globe. She is an avid investigator focused on Knowledge Management Systems and their impact on organizational operations and strategies. Her professional interests include business strategies and process improvement, sales strategies, knowledge management, business intelligence, machine learning, artificial intelligence, and leadership. She received her Ph.D in Organizational Management with a minor in Information Technologies, and holds two Master degrees in Information Sciences from the Federal University of Bahia (UFBa) and University of North Texas. Dr. Merlo has been the recipient of multiple awards for leadership, data and project management, information literacy, diversity and inclusion, and business development. She was also the recipient of multiple scholarships, including Coordenacão de Aperfeiçoamento de Pessoal de Nível Superior (CAPES) for her Master research and Fundação de Amparo à pesquisa do estado da Bahia (Fapesb) for research focusing on digital repositories in a post graduate program at UNIFACS, University in Salvador, Bahia, Brazil. Dr. Merlo is part of scientific committees for Knowledge and Performance Management, Knowledge-Based System, and the Journal of Information and Knowledge Management journals, among others. Dr. Merlo serves on the international advisory board for the International Conference on Knowledge Management (ICKM), having served as co-chair, organizing the 15th, 18th, and 19th annual ICKM Conferences. Dr. Merlo was the Keynote Speaker for the 15th and 18th ICKM Conferences at the University of Santa Catarina, Brazil in 2019, and 2023. In addition, she has a background as a consultant and professional in the field of library management and has built several information units, including libraries and archives. She worked for the City of Austin, Texas for a decade, where she managed multiple projects within Information Sciences, chaired the Culture and Diversity Committee for several years and had an active role serving as chair for the Texas Library Association. Dr. Merlo was the recipient of the Junior Guild Award in 2015 for best project and leadership on diversity and inclusion in public libraries in the state of Texas. Dr. Merlo's research interests include Knowledge Management Systems (tools, technologies, governance, and processes), information systems, artificial intelligence and machine learning, business intelligence, data analytics and governance, leadership, and cloud computing. Her professional experience in the private sector includes work at multiple Fortune 100 companies. She is the editor and author of the book Understanding,

Implementing, and Evaluating Knowledge Management in Business Settings and AI and Data Analytics Applications in Organizational Management by IGI-Global.

**Şule Erdem Tuzlukaya** is the director of Atılım University Graduate School of Social Sciences. Assoc. Prof. Dr. TUZLUKAYA completed his master's degree in the Management Science Department of the University of Northampton and his doctorate degree in the Management and Organization Department of Başkent University. His research focuses on social networks, social capital and complex systems. His major publications include book chapters and articles in the fields of management and organization.

**Sevilay Ulaş** was born in Istanbul. She completed her primary, secondary and high school education in Istanbul. She started her undergraduate education at Eastern Mediterranean University, Faculty of Communication, Department of Public Relations and Advertising. She continued her education at Yeditepe University, Faculty of Communication, Department of Public Relations and Publicity with a master's degree with thesis and at the same department with a PhD degree from Ege University. She is currently a faculty member at the Department of Public Relations and Advertising, Çankaya University.

# Index

## P



## S



## T



## U

## Ensure Quality Research is Introduced to the Academic Community

# Become a Reviewer for IGI Global Authored Book Projects

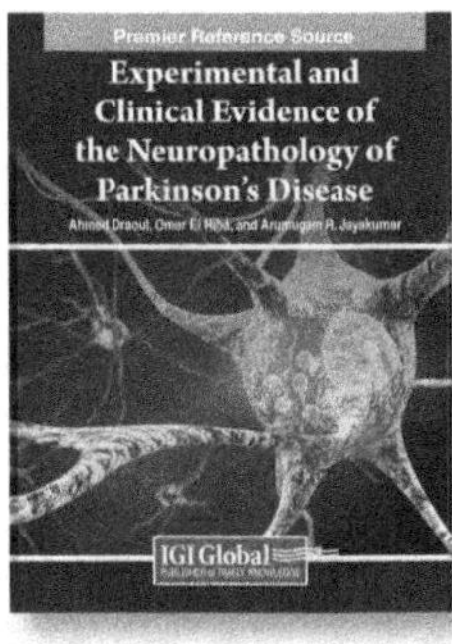

**The overall success of an authored book project is dependent on quality and timely manuscript evaluations.**

**Applications and Inquiries may be sent to:**
development@igi-global.com

Applicants must have a doctorate (or equivalent degree) as well as publishing, research, and reviewing experience. Authored Book Evaluators are appointed for one-year terms and are expected to complete at least three evaluations per term. Upon successful completion of this term, evaluators can be considered for an additional term.

If you have a colleague that may be interested in this opportunity, we encourage you to share this information with them.

www.ingramcontent.com/pod-product-compliance
Lightning Source LLC
LaVergne TN
LVHW081321300926
849161LV00008B/1842
* 9 7 9 8 3 6 9 3 0 7 3 2 8 *